# Student Workbook
(Chapters 1 - 14)

to accompany

# Physics
*A Contemporary Perspective*
Preview Edition

### Randall D. Knight
California Polytechnic State University
San Luis Obispo

Addison-Wesley Publishing Company

Reading, Massachusetts • Menlo Park, California • New York
Don Mills, Ontario • Wokingham, England • Amsterdam • Bonn
Sydney • Singapore • Tokyo • Madrid • San Juan • Milan • Paris

Reproduced by Addison-Wesley from camera-ready copy supplied by the authors.

Copyright © 1996 by Addison-Wesley Publishing Company

All rights reserved. No part of this publication may be reproduced, stored in a retrieval system, or transmitted, in any form or by any means, electronic, mechanical, photocopying, recording, or otherwise, without the prior written permission of the publisher. Printed in the United States of America.

ISBN 0-201-49906-1
1 2 3 4 5 6 7 8 9 10- CRC-99989796

# Table of Contents

1. Concepts of Motion
2. Vectors and Coordinate Systems
3. From Words to Symbols
4. Kinematics: The Mathematics of Motion
5. Force and Motion
6. Dynamics I: Newton's Second Law
7. Dynamics II: Motion in a Plane
8. Dynamics III: Newton's Third Law
9. Momentum and its Conservation
10. Concepts of Energy
11. Work and Energy
12. Potential Energy
13. Expanding the Concept of Energy: Systems of Particles
14. Newton's Law of Gravity

# INTRODUCTION

# A PHYSICS WORKBOOK

Note: This Preview Edition of *A Physics Workbook* contains only Chapters 1 - 14. The workbook will be extended to all chapters when the Preliminary Edition appears in 1996.

*A Physics Workbook* is a companion to the text *Physics: A Contemporary Perspective*. This workbook consists of exercises that give you an opportunity to practice the ideas and techniques presented in the text and in class. These exercises are intended for you to do on a daily basis, right after the topics have been discussed in class and are still fresh in your mind. Learning physics, as in learning any skill, requires *regular practice* of the *basic techniques*, and that is what this workbook is all about. Successful completion of the workbook exercises will prepare you to tackle the more challenging end-of-chapter homework problems in the text.

You will find that nearly all of the exercises are *qualitative* rather than *quantitative*. They ask you to draw pictures, interpret graphs, write short explanations, or provide other answers that do not involve calculations or numbers. There will be ample opportunity, with the end-of-chapter homework, for numbers and problem solving. The purpose of these exercises is to get you thinking along the right directions so that you will have the basic thinking tools *in place* when you get to quantitative problems.

The exercises in this workbook are keyed to *specific sections* of the text. They assume that you have read that section, and they will give you a chance to practice the new ideas introduced in that section. You should keep the text beside you as you work and refer to it often. You will usually find guidelines, figures, or examples in the text that are directly relevant to the exercises. When asked to draw figures or diagrams, you should attempt to draw them so that they look much like figures and diagrams in the text.

Since the exercises relate to specific sections, you should answer them on the basis of information presented in *just* that section. You may have learned new information in Section 7 of a chapter, but you should not use that when answering exercises from Section 4. There will be ample opportunity, when you reach the Section 7 exercises, to use that information there.

You will need some "tools" to complete these exercises. Many of them will ask you to *color code* your answers by drawing some items in black, others in red, and perhaps yet others in blue. You will need to purchase a few colored *pencils* to do this. The author highly recommends that you work in pencil, rather than ink, so as to facilitate erasures and corrections. Few are the individuals who make so few mistakes as to be able to work in ink! In addition, you'll find that a small six-inch ruler, which is easy to carry, will come in handy for drawings and graphs. Finally, it is recommended that, if you don't already have one, you purchase a small stapler. Your instructor will assign certain pages of the workbook to do after each class, and you will need to tear these out, *staple them*, and submit them at the next class meeting.

You will find, as the year goes along, that physics is a way of *thinking* about how the world works and why things happen as they do. We will be interested primarily in finding relationships and seeking explanations, only secondarily in computing numerical answers. In many ways, the "thinking tools" developed in this workbook are what the course is all about. If you take the time to do these exercises regularly *and* to study the graded exercises that are returned, in order to learn from your mistakes, you are guaranteed to do well in this course.

# Chapter 1

# Concepts of Motion

1.1 - 1.3     No Exercises

1.4     The Particle Model

Using the particle model, draw motion diagrams for each motion described below. Number the positions in order, as shown in Fig. 9 in the text. BE NEAT AND ACCURATE!

1. A car accelerates forward from a stop sign and eventually reaches a steady cruising speed of 45 miles per hour.

2. An elevator starts from rest at the 100th floor of the Empire State Building and descends, with no intermediate stops, until coming to rest on the ground floor. (Draw this one *vertically* since the motion is vertical.)

3. Suzy Skier starts *from rest* at the top of a 30° snow-covered slope and skies to the bottom. (Orient your diagram correctly, as seen from the *side*, and label the 30° angle.)

Workbook Chapter 1

4. The Space Shuttle orbits the earth in a circular orbit, completing one revolution each 90 minutes.

5. Bob throws a ball at an upward 45° angle from a third story balcony. The ball lands on the ground below.

Several motion diagrams are shown below. For each, write a short description of the motion of an object that will match the diagram. Your descriptions should name specific objects and be phrased similarly to the descriptions of Exercises 1 - 5. (Note the axis labels on 8 and 9.)

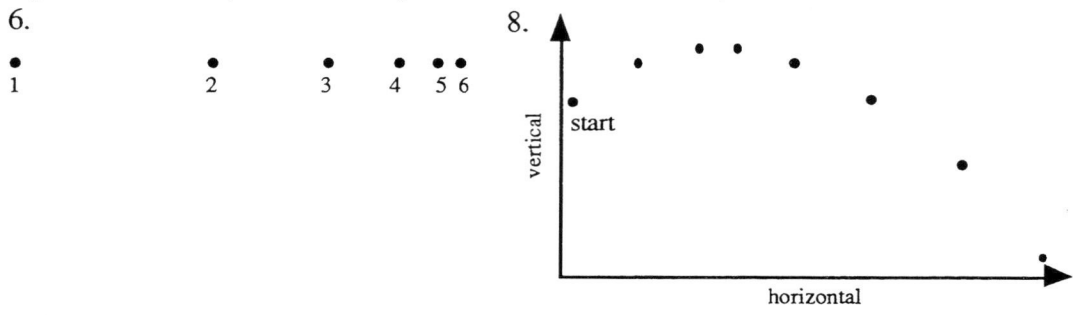

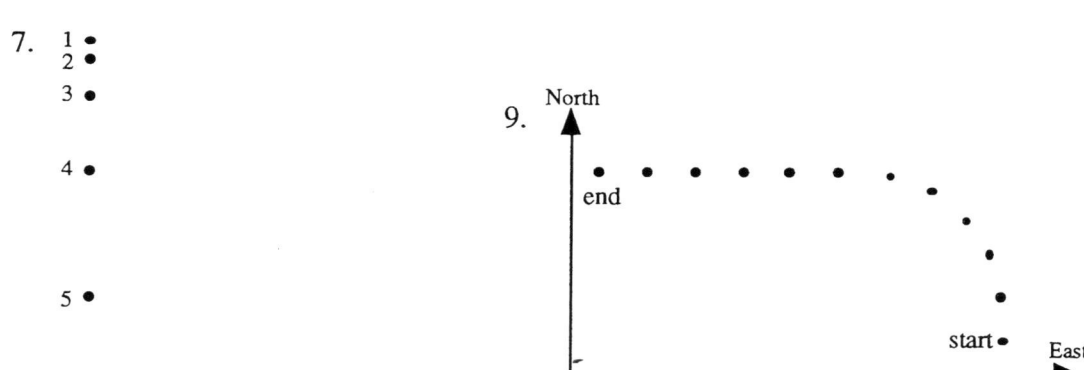

## 1.5 POSITION AND DISPLACEMENT

10. The figure below shows the location of an object at three successive times: 1, 2 and 3.

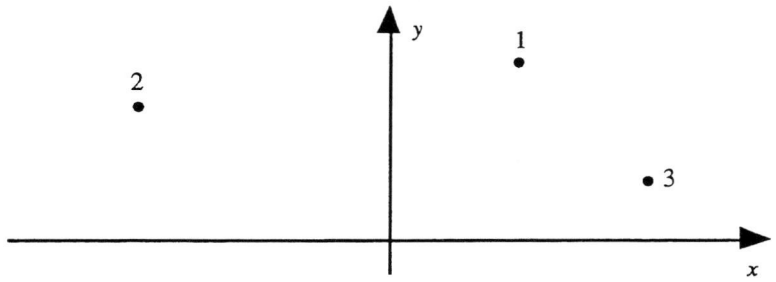

a) Using black pencil: draw *and label*, on the figure, the three position vectors $\vec{r}_1$, $\vec{r}_2$, and $\vec{r}_3$ at times 1, 2, and 3.
b) Using blue or green pencil: Draw the object's "trajectory" 1 → 2 → 3.
c) Using red pencil: Draw the displacement vector $\Delta \vec{r}$ from the initial to the final position.

11. Referring to question 10, is the above object's displacement equal to the distance the object travels? Explain.

12. Redraw your five motion diagrams of Exercises 1 - 5 in the space below, but now add and label the displacement vectors $\Delta \vec{r}$ on the diagrams.

Workbook Chapter 1

## 1.6   VELOCITY

13. The figure below shows the positions of a moving object in two successive frames of film. Frame 1 occurs prior to frame 2.

2 •

• 1

a)   Using red pencil, draw the velocity vector corresponding to these two frames.
b)   Establish a coordinate system of your choice by drawing a set of *xy*-axes. Place the origin wherever you wish. Then, using black pencil, draw *and label* the position vectors $\vec{r}_1$ and $\vec{r}_2$.
c)   Does $\vec{r}_2 = \vec{r}_1 + \vec{v}$ on *your diagram*? Explain.

Draw motion diagrams for the following motions (Exercises 14 - 17). Use *at least* six points for each. more if necessary to make the motion clear. Show and label the *velocity* vectors.

14. A "rocket car" accelerates from rest to a high speed, then coasts at constant speed after running out of fuel. Use a dotted line across your diagram to indicate the point at which the car runs out of fuel.

15. Galileo drops a ball from the Leaning Tower of Pisa. Consider its motion from the moment it leaves his hand until a microsecond before it hits the ground. You diagram should be vertical.

Workbook Chapter 1

16. An elevator starts from rest on the ground floor, accelerates upward for a short time, then moves with constant speed, and finally brakes to a halt at the tenth floor. Use dotted lines across your diagram to indicate where the acceleration stops and where the braking begins. You'll need ten or more points on this one to indicate the motion clearly.

17. A bowling ball being returned from the pin area to the bowler rolls at a constant speed, then up a ramp, and finally exits onto a level section at very low speed. You'll need ten or more points on this one to indicate the motion clearly.

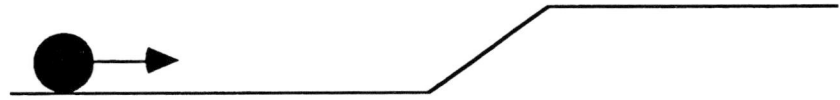

18. Tommy Trackstar runs once around the usual running track (straight sides with semi-circular ends) at constant speed. Include about 20 points on your motion diagram.

Workbook Chapter 1

## 1.7 ACCELERATION

**NOTE:** Beginning with this section, and for all future motion diagrams, we will "color code" the vectors. You should draw velocity vectors with *black* pencil and acceleration vectors with *red* pencil.

The figures below show an object's position in three successive frames of film. The object is moving in the direction $1 \to 2 \to 3$. For each diagram:

a) Draw *and label* ($\vec{v}_i$ and $\vec{v}_f$) the velocity vectors on the motion diagram given.
b) Below the motion diagram, redraw the two velocity vectors with their tails together. This will look similar to Figure 19b in the text.
c) Determine the acceleration $\vec{a}$ from $\vec{v}_i$ and $\vec{v}_f$, using vector subtraction
d) Draw and label $\vec{a}$ at the proper location on the motion diagram, similar to Fig. 19c.
e) Decide whether the object is speeding up, slowing down, or moving at a constant speed. Write your answer beside the diagram.

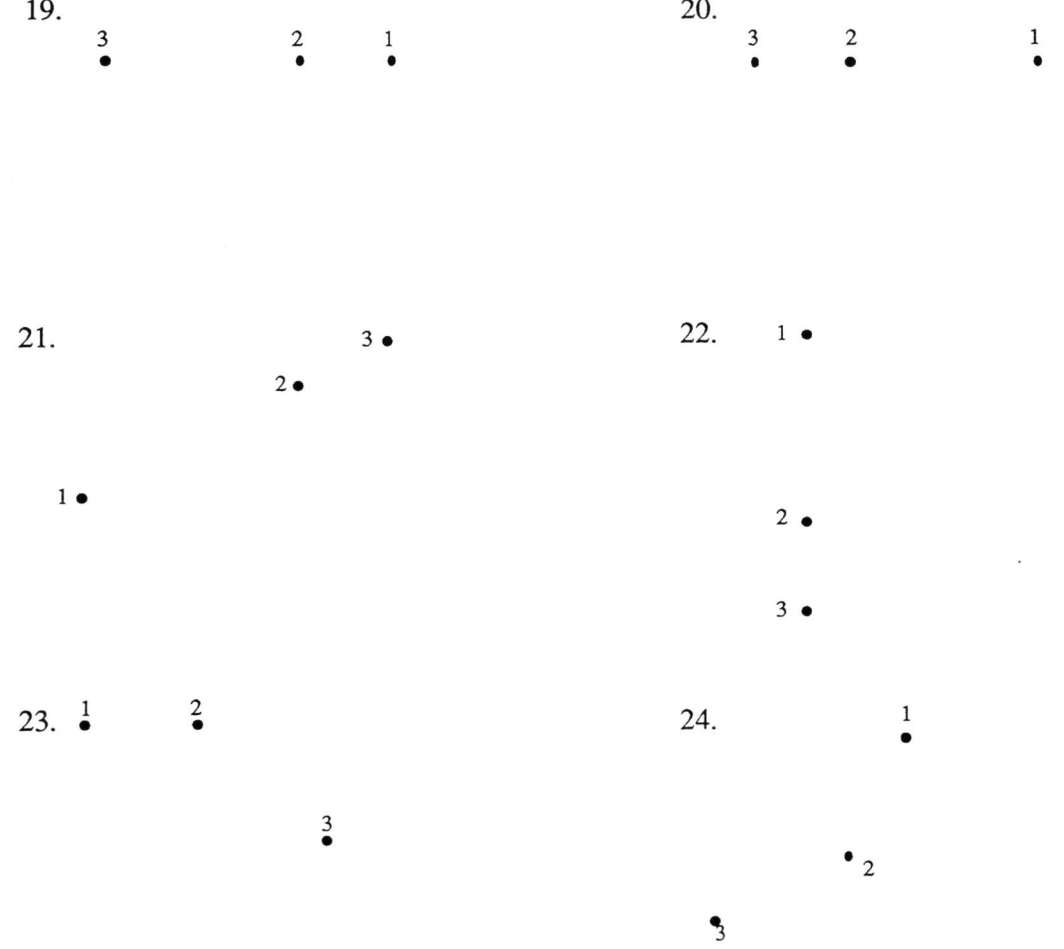

Workbook Chapter 1

Draw complete motion diagrams for Exercises 25 - 30. Include and label velocity vectors $\vec{v}$ and acceleration vectors $\vec{a}$. Color code them appropriately.

25. Galileo drops a ball from the Leaning Tower if Pisa. Consider its motion from the moment it leaves his hand until a microsecond before it hits the ground.

26. Betty is driving her car at a steady 30 mph when a small furry creature runs into the road in front of her. She hits the brakes and skids to a stop. Consider her motion from 1 second before braking until she has stopped.

27. A ball is rolled up a smooth board tilted at a 30° angle, and then it rolls back to its starting position.

28. A bowling ball being returned from the pin area to the bowler rolls at a constant speed, then up a ramp, and finally exits onto a level section at very low speed.

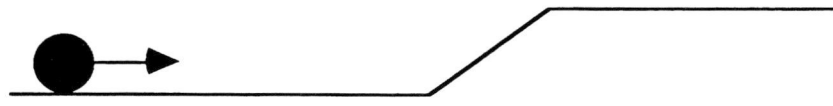

29. Tommy Trackstar runs once around the usual running track (straight sides with semi-circular ends) at constant speed. Include about 20 points on your motion diagram.

30. A cannon ball is fired from a Civil War cannon up onto a high cliff. Consider the cannon ball from the moment it leaves the cannon until a microsecond before it hits the ground.

Workbook Chapter 1

1.8 - 1.9    No Exercises

# Chapter 2

# Vectors and Coordinate Systems

2.1 - 2.2    No Exercises

2.3    Vector Arithmetic

Find and label the vector sum $\vec{A} + \vec{B}$ for the vectors shown below.

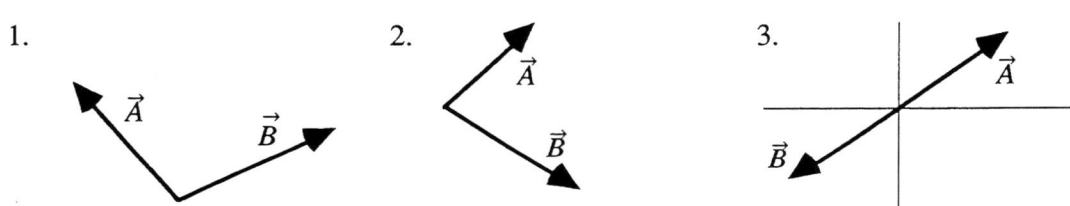

4. Use a figure (similar to Fig. 6) to show that vector addition is associative -- that is, that
$$(\vec{A} + \vec{B}) + \vec{C} = \vec{A} + (\vec{B} + \vec{C}).$$

Find and label the vector difference $\vec{A} - \vec{B}$ for the vectors shown below.

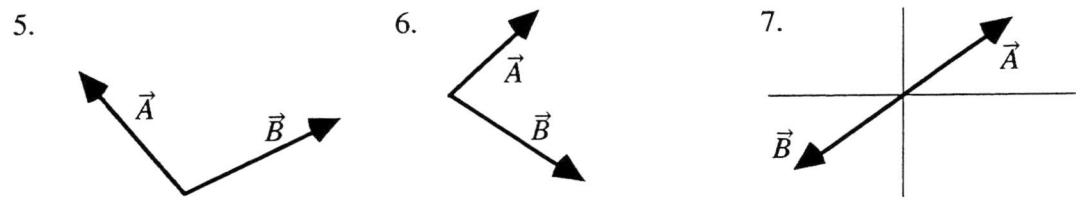

Workbook Chapter 2

8. Draw and label the vector $2\vec{A}$ and the vector $\frac{1}{2}\vec{A}$.

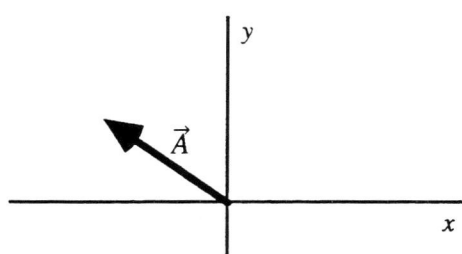

9. Is it possible to <u>add</u> a scalar to a vector? If so, demonstrate. If not, explain why.

10. How would you define the "zero vector" $\vec{Z} = 0$?

## 2.4  COORDINATE SYSTEMS AND COMPONENT VECTORS

Draw and label the *x*- and *y*-component vectors for each of the following:

11.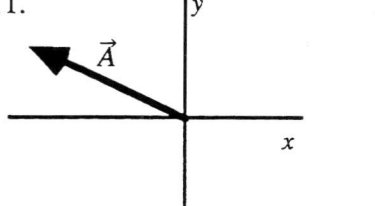

12.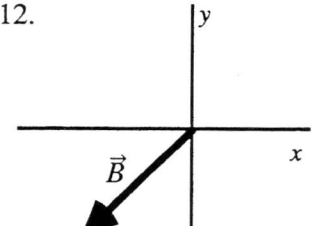

13.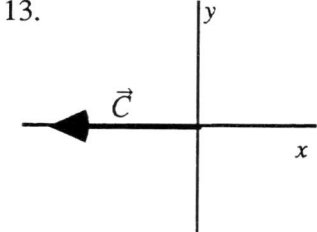

## 2.5  UNIT VECTORS AND VECTOR COMPONENTS

Write the following vectors in component form (e.g. $3\hat{i} + 2\hat{j}$).

14. 

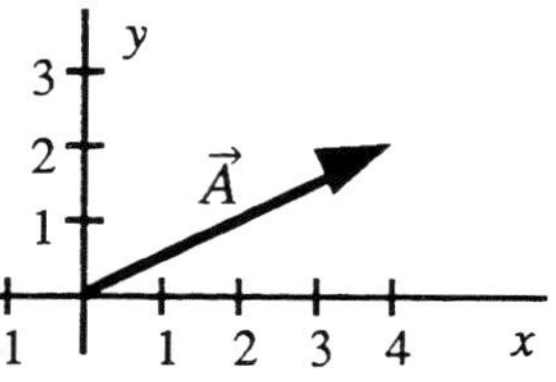

15.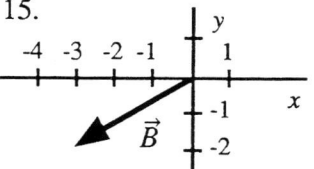

16.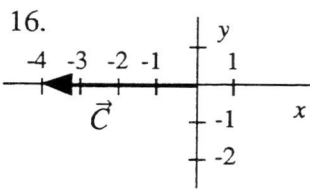

Workbook Chapter 2

17. What is the vector sum $\vec{D} = \vec{A} + \vec{B} + \vec{C}$ for the three vectors (Ex. 14-16) above? Write your answer in *component* form.

Draw and label the following vectors on the axes below:

18. $-\hat{i} + 2\hat{j}$

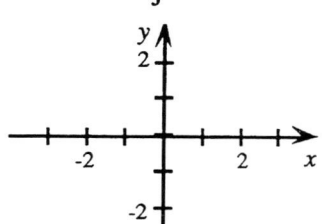

19. $-2\hat{j}$

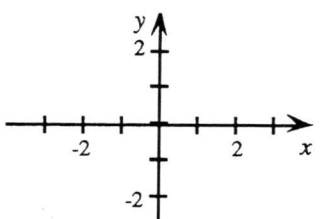

20. $3\hat{i} - 2\hat{j}$

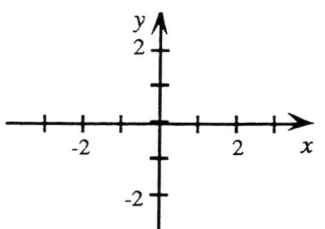

21. Using the idea of components, how would you define the "zero vector" $\vec{Z} = 0$?

## 2.6   MOVING BETWEEN THE TWO VIEWPOINTS

Compute numerical values for the *x*- and *y*-components of these vectors:

22.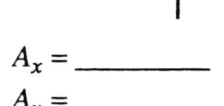

$A_x =$ _____
$A_y =$ _____

23.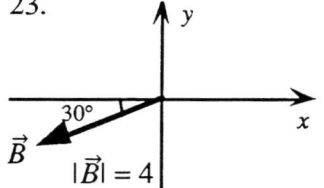

$B_x =$ _____
$B_y =$ _____

24.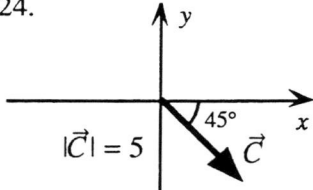

$C_x =$ _____
$C_y =$ _____

2-3

Workbook Chapter 2

For each of the three vectors below:
   a) Draw the vector on the axes provided.
   b) Identify and label an angle $\theta$ describing the direction of the vector.
   c) Find the magnitude and the angle of the vector.

25. $\vec{A} = 2\hat{i} + 2\hat{j}$

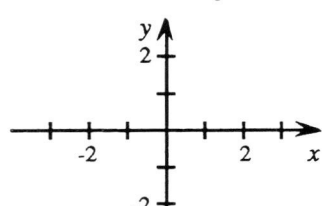

$|\vec{A}| = $ _____

$\theta = $ _____

26. $\vec{B} = -2\hat{i} + 2\hat{j}$

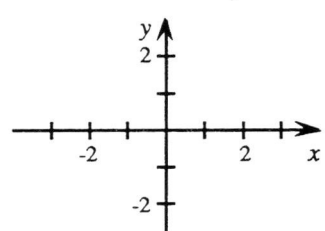

$|\vec{B}| = $ _____

$\theta = $ _____

27. $\vec{C} = -3\hat{i} - 2\hat{j}$

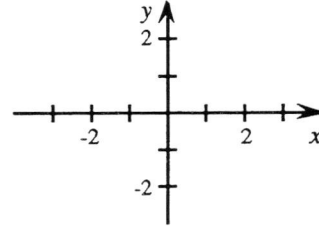

$|\vec{C}| = $ _____

$\theta = $ _____

28. Can a vector have a component equal to zero and still have nonzero magnitude? Explain.

29. Can a vector have zero magnitude if one of its components is nonzero? Explain.

30. Two vectors have unequal magnitudes. Can their sum be zero? Explain.

2 - 4

## 2.7 TILTED AXES AND ARBITRARY DIRECTIONS

Vector $\vec{A}$ is defined as $\vec{A} = (5, 30°$ above the horizontal$)$. Find the components $A_x$ and $A_y$ in the three coordinate systems shown below. Show your work below the figure.

31.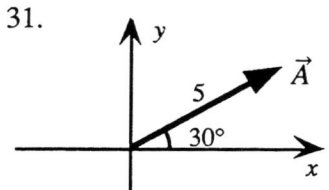

$A_x =$ _____
$A_y =$ _____

32.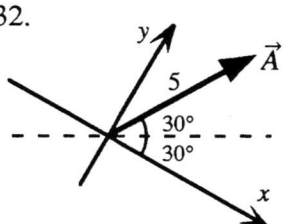

$A_x =$ _____
$A_y =$ _____

33.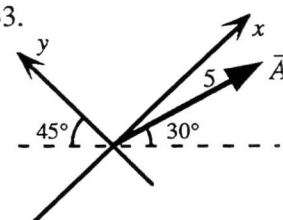

$A_x =$ _____
$A_y =$ _____

## 2.8 SIGNIFICANT FIGURES

34. How many significant figures does each of the following numbers have?

   a) 6.21  _____
   b) 62.1  _____
   c) 6210  _____
   d) 62100 _____
   e) 62100. _____
   f) 0.0621 _____
   g) 0.620 _____
   h) 0.62  _____
   i) .62   _____
   j) 1.0621 _____
   k) $6.21 \times 10^3$ _____
   l) $62.1 \times 10^3$ _____

35. Compute the following numbers, applying the significant figure standard adopted for this course.

   a) $33.3 \times 25.4 =$ _____
   b) $33.3 - 25.4 =$ _____
   c) $33.3 \div 25.4 =$ _____
   d) $33.3 \times 45.1 =$ _____
   e) $33.3^2 =$ _____
   f) $33.3^{1/2} =$ _____

Workbook Chapter 2

# Chapter 3

# From Words To Symbols

3.1 - 3.2    No Exercises

3.3    The Pictorial Model

1.  Summarize the major steps in an "expert's" approach to problem solving.

2.  You probably have never thought about your own "strategy" for problem solving, but now is a good time to reflect on how <u>you</u> approach a new problem. Spend a few minutes thinking about it, then summarize the major steps in your approach to problem solving.

3.  By comparing your answers to 1 and 2:
a)    Can you identify any "weaknesses" in your problem solving strategy that you need to work on avoiding.

Workbook Chapter 3

b) Can you identify new tactics that you should practice adding to your strategy?

## 3.5   USING SYMBOLS

4. The motion diagrams below show an initial point 0 and a final point 1. A pictorial model would define five symbols: $x_0$ and $x_1$ (or $y_0$ and $y_1$), $v_0$ and $v_1$, and $a$. Determine whether each of these quantities is positive, negative, or zero. Place either +, –, or 0 in each cell of the table below.

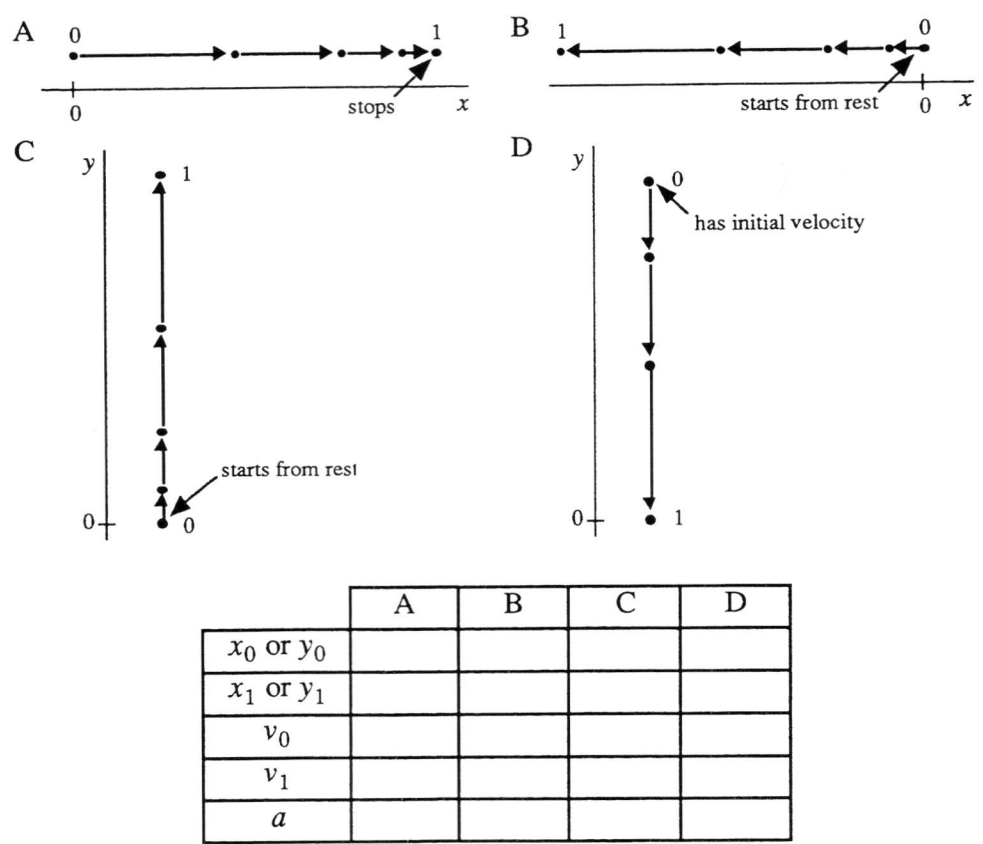

|  | A | B | C | D |
|---|---|---|---|---|
| $x_0$ or $y_0$ |  |  |  |  |
| $x_1$ or $y_1$ |  |  |  |  |
| $v_0$ |  |  |  |  |
| $v_1$ |  |  |  |  |
| $a$ |  |  |  |  |

Workbook Chapter 3

5. The three symbols $x$, $v$, and $a$ have eight possible combinations of *signs*. For example, one combination is $(x, v, a) = (+, -, +)$.

a) List all eight combinations.

| # | combination | # | combination |
|---|---|---|---|
| 1 | _____ | 5 | _____ |
| 2 | _____ | 6 | _____ |
| 3 | _____ | 7 | _____ |
| 4 | _____ | 8 | _____ |

b) For each of the eight combinations of signs you identified in Part a), draw a complete motion diagram showing the motion of an object having these signs for $x$, $v$, and $a$. Draw the diagram *below* the axis whose number corresponds to Part a), and use labeled and color-coded $\vec{v}$ and $\vec{a}$ vectors.

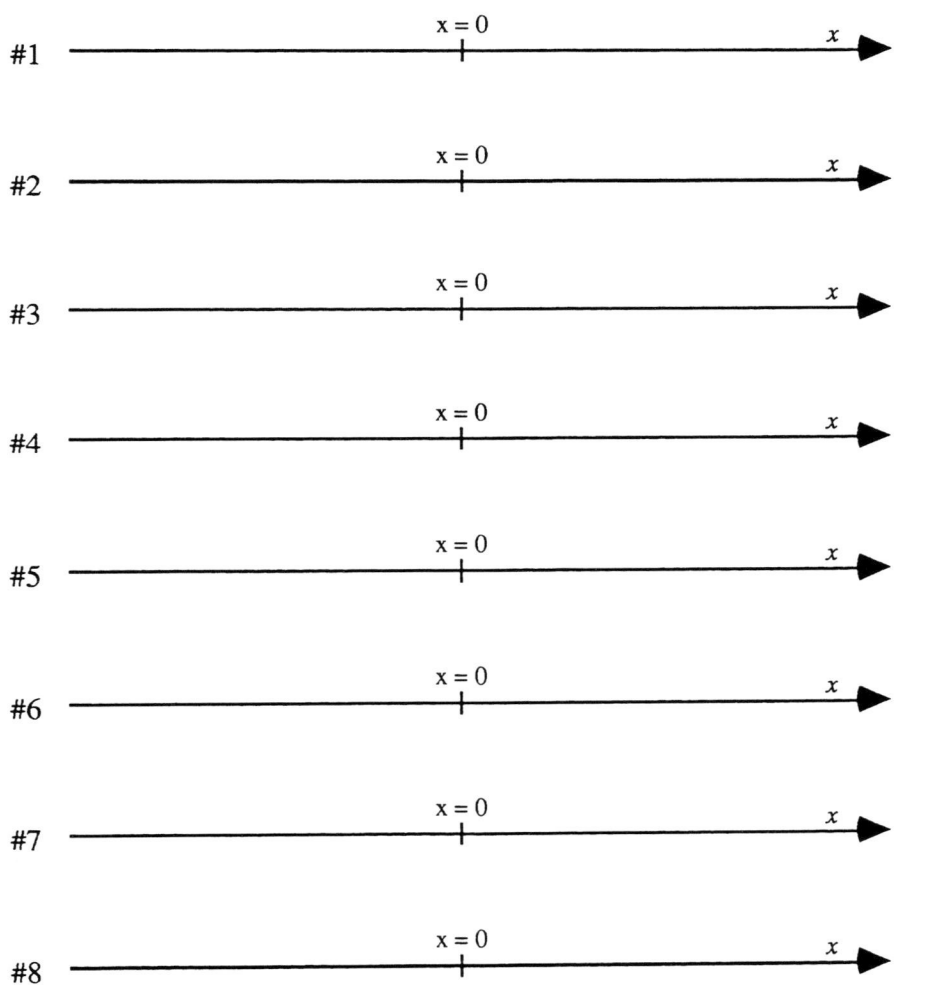

3.6    No Exercises

# CHAPTER 4

# KINEMATICS

4.1 - 4.2     No Exercises

4.3     MEASUREMENTS AND UNITS

1. Convert the following to SI units:
    a) 9.12 µs           b) 3.42 km          c) 44 cm/ms         d) 80 km/hour
    e) 250 cm$^3$        f) 8 inches         g) 14 square inches h) 60 miles/hour
    [Note: Think carefully about e) and g).  A geometric picture may help.]

Workbook Chapter 4

2. Use Table 4-3 to assess whether or not the following statements are *reasonable*. Note the comments on *assessment* and *reasonableness* at the end of Section 4.3.
   a) Joe is 180 cm tall.

   b) I rode my bike to campus at a speed of 50 m/s.

   c) Sammy Skier reaches the bottom of the slope going 25 m/s.

   d) I can throw a ball 2 km.

   e) I can throw a ball with a speed of 100 km/hour.

3. Justify the assertion that 1 m/s ≈ 2 mph by *exactly* converting 2 mph to SI units. By what percentage is this rough conversion in error?

## 4.4  POSITION AND TIME

4. Sketch position-versus-time graphs for the following motions. Include a numerical scale on both axes, with units that are *reasonable* for this motion. Some numerical information is given in the problem, but for other quantities make *reasonable estimates*. [Note: A *sketched* graph simply means hand-drawn, rather than carefully measured and laid out with a ruler. But a sketch should still be neat and as accurate as is feasible by hand; it also should include labeled axes and, if appropriate, it should have tick-marks and numerical scales along the axes.]

a) A student walks to the bus stop, waits for the bus, then rides to campus. Assume that all the motion is along a straight street.

b) A students walks slowly to the bus stop, realizes he forgot his paper that is due, and *quickly* walks home to get it.

c) The quarterback drops back 10 yards from the line of scrimmage then throws a pass 20 yards to the tight end, who catches it and sprints 20 yards to the goal. Draw your graph for the *football*. Think carefully about what the *slopes* of the lines should be.

d) Follow the horizontal motion of a tennis ball through several volleys back and forth. Let the net be at $x = 0$ and the server be on the left side of the net.

Workbook Chapter 4

5. Interpret the following position-versus-time graphs by writing a very short "story" of what is happening. Be creative! Have characters and situations. Simply saying that "a car moves 100 meters to the right" is too dull to qualify as a story. Your stories should make *specific reference* to information obtained from the graphs, such as distances moved or time elapsed.

a) Moving car.

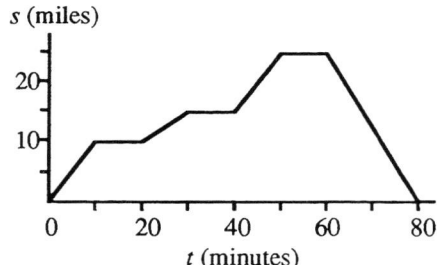

b) Sprinter.

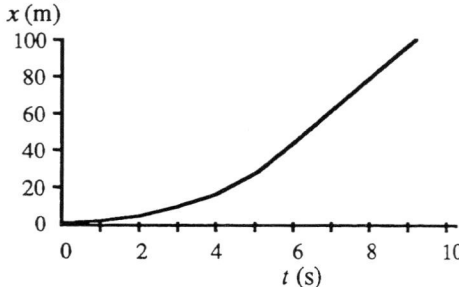

c) Submarine.

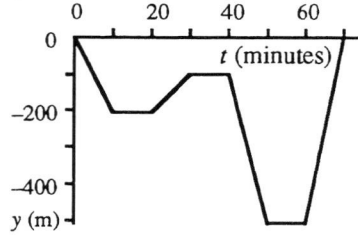

d) Two football players.

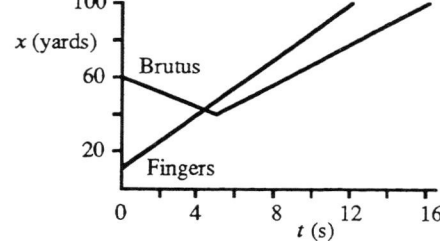

4 - 4

6. Consider this position-versus-time graph, which shows a *single* curve. Can you give this graph an interpretation? If so, then do so. If not, why not?

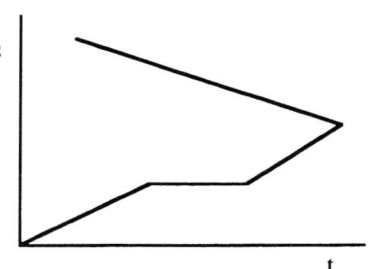

## 4.5    UNIFORM·MOTION

7. Sketch position-versus-time graphs for the following motions. Include appropriate numerical scales along both axes. A small amount of computation is necessary.

   a) A parachutist open her parachute at an altitude of 1500 m, and she then descends slowly to earth at a steady speed of 5 m/s. Start your graph as her parachute opens.

   b) Trucker Bob starts the day 100 miles west of Denver. He drives eastward for 3 hours at a steady 60 miles/hour before stopping for his coffee break. Let Denver be located at $x = 0$ and assume that the $x$-axis points to the east.

   c) Quarterback Bill passes the ball straight down field with a speed of 15 m/s. It is intercepted 45 m away by Linebacker Larry, who is running up field at 7.5 m/s. He carries it 60 m to score. Draw the graph for the *football*.

Workbook Chapter 4

8. The figure shows a position-versus-time graph for the motion of two objects, A and B, that are moving along the same axis.
   a) At the instant $t = 1$ s, is the speed of A greater than, less than, or equal to the speed of B? Explain your reasoning.

   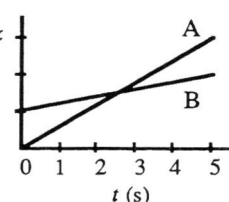

   b) Do object's A and B ever have the *same* speed? If so, at what time or times? Explain your reasoning.

9. Interpret the following position-versus-time graphs by writing a very short "story" of what is happening. Your stories should make *specific reference* to the *speed* of the moving objects, which you will need to determine from the graphs. Assume that the motion takes place along a horizontal line.

   (a)
   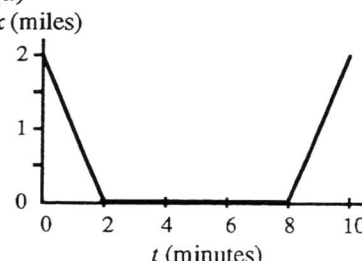

   (b)
   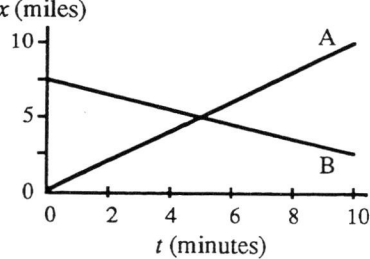

   (c)
   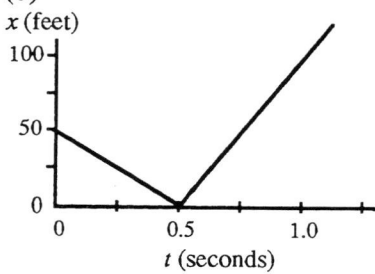

## 4.6  INSTANTANEOUS VELOCITY

10. Draw both a position-versus-time graph and a velocity-versus time graph for an object at rest at $x = 1$ m.

11. Below are six position-versus-time graphs. For each, draw the corresponding velocity-versus-time graph directly below it. Both graphs should have the same time scale (i.e., a vertical line drawn through both graphs would connect the velocity $v$ at time $t$ with the position $s$ at the same time $t$). While there are no numbers in this question, your graphs should indicate the *relative* speeds correctly.

a)
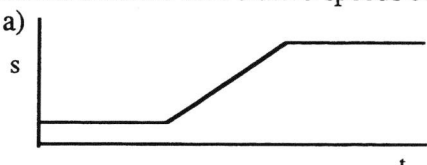

b)
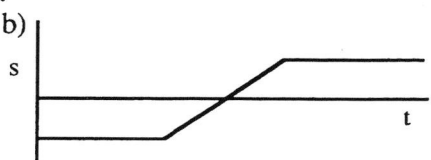

c)
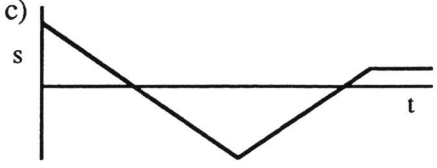

d)
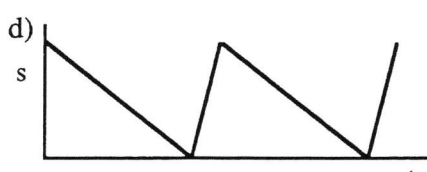

Workbook Chapter 4

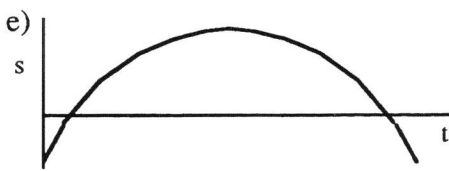

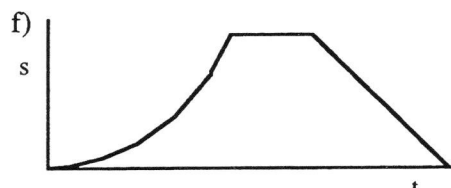

12. The figure shows a position-versus-time graph for the motion of two objects, A and B, that are moving along the same axis.

    a) At the instant $t = 1$ s, is the speed of A greater than, less than, or equal to the speed of B? Explain your reasoning.

    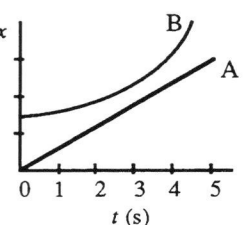

    b) Do objects A and B ever have the *same* speed? If so, at what time or times? Explain your reasoning?

13. The figure shows a position-versus-time graph for a moving object. Various specific times are labeled A, B, ..., E. At which lettered point or points:

    a) Is the motion slowest?

    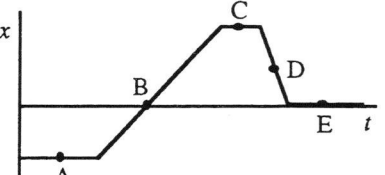

    b) Is the motion the fastest?

    c) Is the object moving at a constant non-zero velocity?

    d) Is the object moving to the left?

4 - 8

e) Is the object turning around?

f) Is the object at rest?

14. The figure shows a position-versus-time graph for a moving object. Various specific times are labeled A, B, ..., F. At which lettered point or points:
   a) Is the motion slowest?

   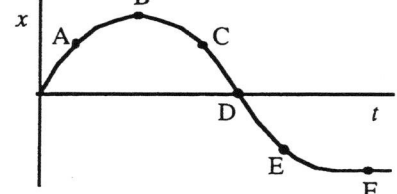

   b) Is the motion the fastest?

   c) Is the object moving at a constant non-zero velocity?

   d) Is the object moving to the left?

   e) Is the object speeding up?

   f) Is the object slowing down?

   g) Is the object turning around?

15. For each of the following, draw
   i) A motion diagram,
   ii) A position-versus-time graph, and
   iii) A velocity-versus-time graph.
   a) A car starts from rest, steadily speeds up to 40 mph in 15 s, moves at a constant speed for 30 s, then brakes rapidly to a halt in 5 s.

Workbook Chapter 4

b) A rock is dropped from a bridge and steadily speeds up to 30 m/s when it hits the ground 3 s later. Let the y-axis point vertically up, and think carefully about the signs.

c) A pitcher winds up and throws a baseball with a speed of 40 m/s. One-half second later the batter hits a line drive with a speed of 60 m/s. The ball is caught 1 s after it is hit. From where you are sitting, the batter is to the right of the pitcher. Draw your motion diagram and graph for the *ball*.

16. The figure shows six frames from the motion diagram of two moving cars, A and B.
    a) Describe, in words, what you would *see* if you watched this motion.

b) Draw both a position-versus-time graph and a velocity-versus-time graph. Show the motion of both cars on each graph, such as in Exercises 8 or 12, and label them A and B.

4 - 10

c) Do the two cars ever have the same position at one instant of time? If so: i) identify in which frame number, and b) identify the point on your graphs of part b).

d) Do the two cars ever have the same velocity at one instant of time? If so: i) identify in which frame number, and b) identify the point on your graphs of part b).

17. The figure shows six frames from the motion diagram of two moving cars, A and B.

   a) Describe, in words, what you would *see* if you watched this motion.

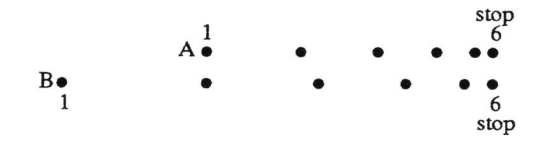

   b) Draw both a position-versus-time graph and a velocity-versus-time graph. Show the motion of both cars on each graph, such as in Exercises 8 or 12, and label them A and B.

   c) Do the two cars ever have the same position at one instant of time? If so: i) identify in which frame number, and b) identify the point on your graphs of part b).

   d) Do the two cars ever have the same velocity at one instant of time? If so: i) identify in which frame number, and b) identify the point on your graphs of part b).

Workbook Chapter 4

18. As you drive along the highway at a steady speed of 60 mph, you notice another car following you. The other driver patiently maintains a constant distance behind you for awhile, but eventually he decides to pass you. At the moment when the front of his car is exactly even with the front of your car, and you turn your head to smile at him, do the two cars have equal positions along the road axis? [They are certainly different in the sense that they are side-by-side. The question is about their position as measured along the road.] Equal velocity? Both? Neither? Explain your reasoning.

## 4.7  RELATING VELOCITY TO POSITION

19. Below are shown two velocity-versus-time graphs. For each:
    i) Draw the corresponding position-versus-time graph,
    ii) Draw a motion diagram, and
    iii) Give a verbal description of the motion.
    Assume that the motion takes place along a horizontal line and that $x(t=0) = 0$.

a)

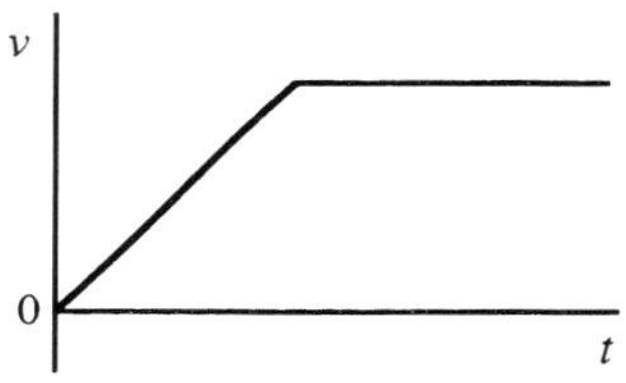

b)

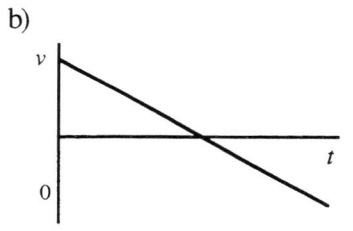

20. The figure shows the velocity-versus-time graph for a moving object whose initial position is $x(t=0) = 20$ m. Find the object's position graphically, using the geometry of the graph, at the following times.

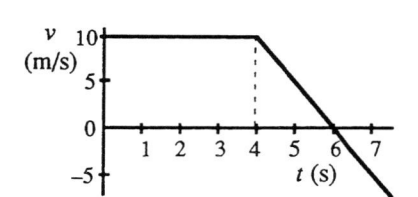

   a) At $t = 3$ s.

   b) At $t = 5$ s.

   c) At $t = 7$ s.

   d) You should have found a very simple relationship between your answers to b) and c). Can you explain this? What is the object doing?

21. Below are shown three velocity-versus-time graphs. Draw the corresponding position-versus-time graphs directly below them. For each, $x(t=0) = 0$.

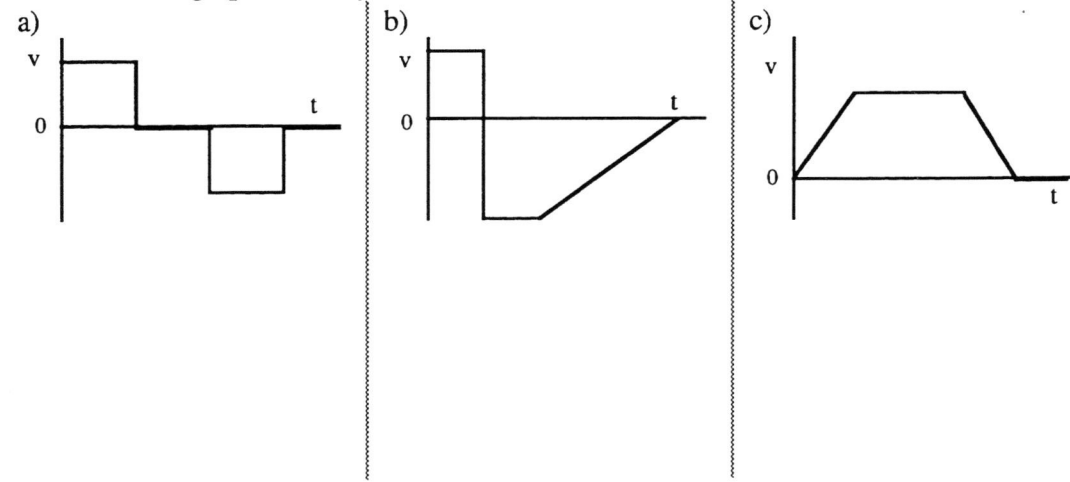

   a)

   b)

   c)

Workbook Chapter 4

Workbook Chapter 4

## 4.8  ACCELERATION

22. A car is traveling north. Can its acceleration vector ever point south? Explain.

23. Give a specific example for each of the following situations. For each, provide:
    i) A description in words, and
    ii) A motion diagram.
    a) $a = 0$ but $v \neq 0$.

    b) $v = 0$ but $a \neq 0$.

    c) $v > 0$ and $a > 0$.

    d) $v > 0$ and $a < 0$.

24. Below are three velocity-versus-time graphs. For each:
    i) Draw, directly below it, the corresponding acceleration-versus-time graph.
    ii) Give a description in words of the motion.

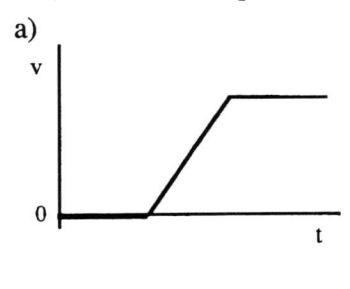

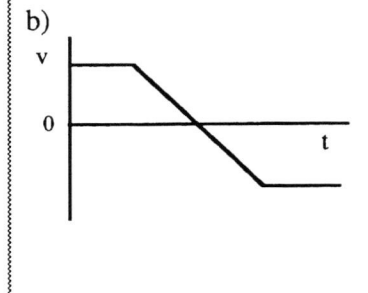

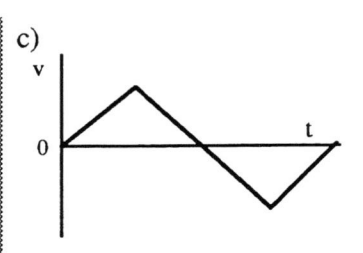

a)

b)

c)

4 - 15

Workbook Chapter 4

25. Below are three acceleration-versus-time curves. For each:
   i) Draw the corresponding velocity-versus-time curve, assuming that $v(t=0) = 0$, and
   ii) Draw a motion diagram.

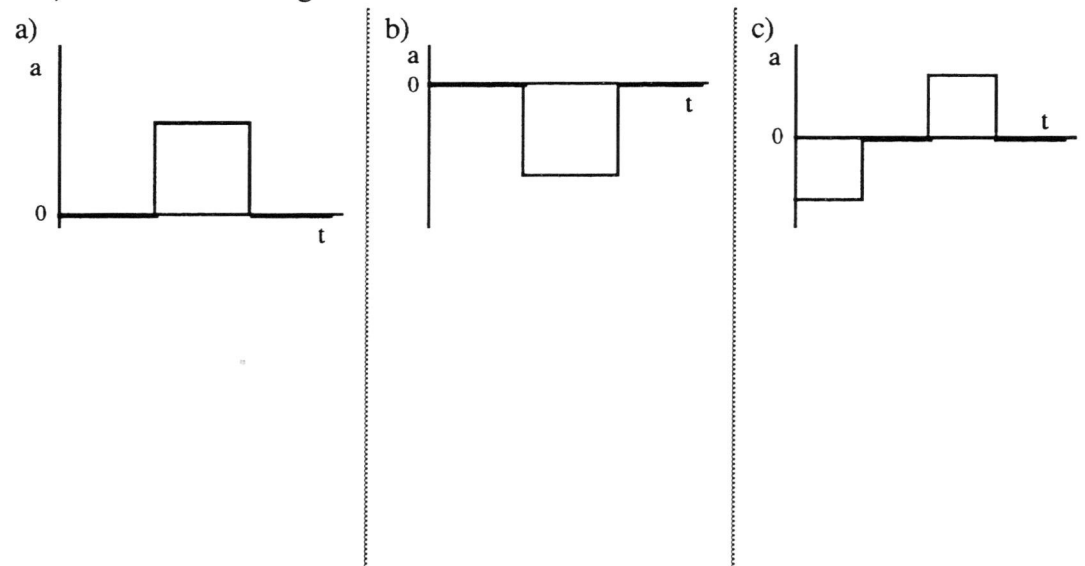

   a)
   b)
   c)

26. The figure below shows 9 frames from the motion diagram of two cars. Both cars begin to accelerate, with constant acceleration, in frame #4.

   a) Which car has the largest initial velocity? How can you tell?

   b) Which car has the largest final velocity? How can you tell?

   c) Which car has the largest acceleration after frame #4? How can you tell?

Workbook Chapter 4

d) Draw position, velocity, and acceleration graphs, showing both cars motion on each graph. This is a total of three graphs with two curves on each.

e) Do the cars ever have the same position at one instant of time? If so, identify in which frame or frames and identify it on your graphs?

f) Do the two cars ever have the same velocity at one instant of time? If so, identify the *two* frames between which this velocity occurs *and* identify it on your graphs.

g) Give a description, in words, of how these two cars move.

4.9    FREE FALL

27. A ball rolling released from rest on an inclined plane has an acceleration of 2 m/s². Complete the table below showing its velocity at the times indicated. Do NOT use a calculator for this -- this is a reasoning question, not a numerical problem.

| time (s) | velocity (m/s) |
|----------|----------------|
| 0        | 0              |
| 1        |                |
| 2        |                |
| 3        |                |
| 4        |                |
| 5        |                |
| 6        |                |

4 - 17

Workbook Chapter 4

28. A ball is thrown straight up into the air. What is the ball's acceleration
   a) Just after leaving your hand?

   b) At the very top (maximum height)?

   c) Just before hitting the ground?

29. Alice throws a red ball straight up into the air, releasing it with velocity $v_0$. As she is throwing it, you happen to pass by in an elevator that is rising with the same constant velocity $v_0$. At the exact instant Alice releases her ball, you reach out of the elevator's window [this is a very fancy elevator!] and *gently* release a blue ball. Both balls are the same height above the ground at the moment they are released.
   a) Describe, in words, the motion of the two balls as Alice sees them from the ground. In what ways are the motion of the red ball and the blue ball the same or different?

   b) Describe the motion of the two balls as you see them from the moving elevator. In what ways are the motion the same or different?

   c) Alice sees a definite "top" of the motion, where her red ball reaches a maximum height and then starts to fall. Call the time of maximum height $t_1$. As you watch from the elevator, do *you* see anything different or unusual about the red ball's motion at $t_1$?

d) Does the red ball "stop" at time $t_1$ when Alice sees it at the very top of its trajectory? As part of answering this question, define what you mean by the word "stop."

30. A rock is *thrown* (not dropped) straight down from a bridge into the river below.
a) Immediately after being released, is the magnitude of the rock's acceleration greater than $g$, less than $g$, or equal to $g$. Explain your reasoning.

b) Immediately before hitting the water, is the magnitude of the rock's acceleration greater than $g$, less than $g$, or equal to $g$. Explain.

… Workbook Chapter 4

# CHAPTER 5

# FORCE AND MOTION

5.1 - 5.2    No Exercises

5.3    FORCE AND MOTION

1. The graph shows a force-versus-acceleration graph, with data recorded for a single object of mass $m$ at several force strengths. Draw and label, directly on these axes, the force-versus-acceleration graphs for objects of mass
   a) $2m$
   b) $0.5m$

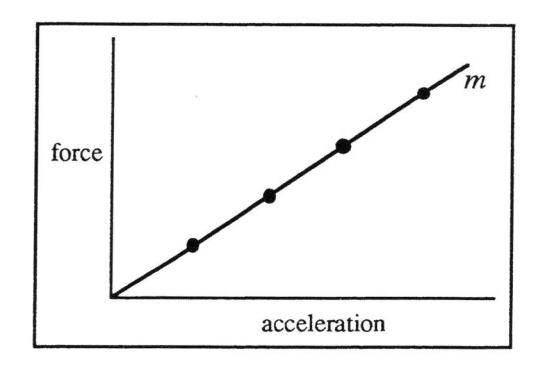

2. A constant force applied to object A produces an acceleration of 5 m/s². The same force applied to object B produces an acceleration of 3 m/s², and applied to object C it produces an acceleration of 8 m/s².
   a) Which object has the largest mass? _____
   b) Which object has the smallest mass? _____
   c) What is the ratio of mass A to mass B ($m_A/m_B$)? _____

3. Based on information presented <u>thus far</u>, what are the *units* of mass? Note: The answer is <u>not</u> kilograms. Read the question carefully.

   What might be the practical difficulties with using these units?

5 - 1

Workbook Chapter 5

4. An object experiencing a constant force undergoes an acceleration of 10 m/s². What will the acceleration of this object be if
   a) The force is doubled? _____   d) The mass is doubled? _____
   b) The force is tripled? _____   e) The mass is tripled? _____
   c) The force is halved? _____   f) The mass is halved? _____
   g) The force and the mass are both doubled? _____
   h) The force is doubled and the mass is halved? _____
   i) The force is halved and the mass is doubled? _____

## 5.4    COMBINING FORCES

5. Forces are shown on three objects below. For each:
   a) Working directly on the figure, determine and show the net force vector.
   b) Show the acceleration vector with a labeled arrow directly below the figure.

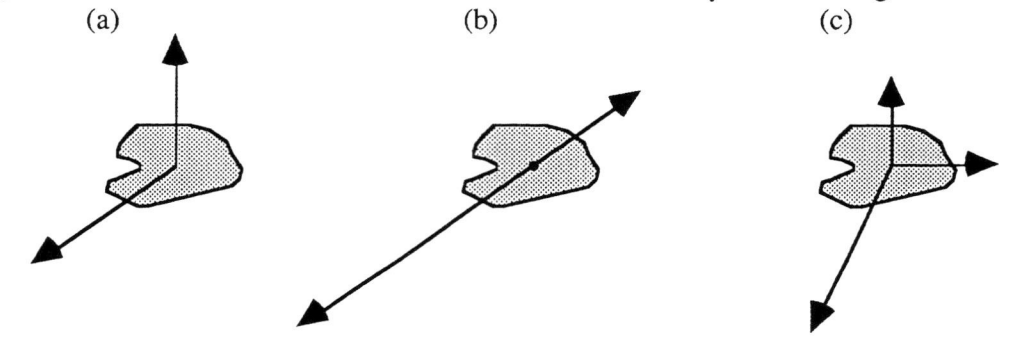

   (a)                (b)                (c)

6. In the figures below, one force is missing. Use the given direction of acceleration to determine the missing force and show it on the figure. Do all work directly on the figure.

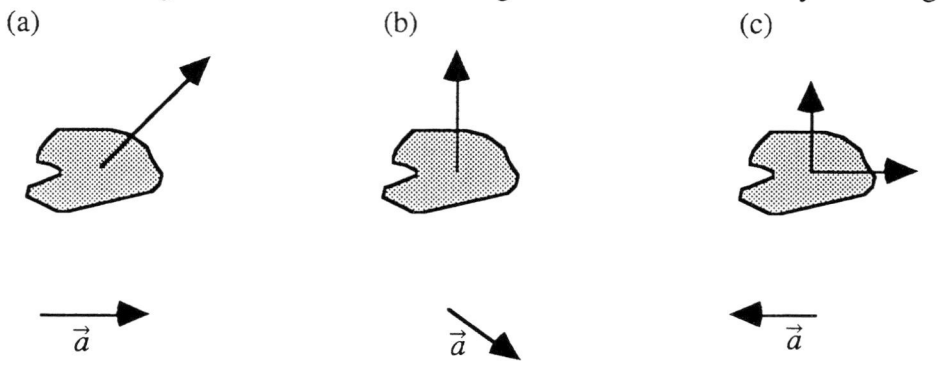

   (a)                (b)                (c)

   $\vec{a}$          $\vec{a}$          $\vec{a}$

## 5.5 INERTIA AND NEWTON'S FIRST LAW

7. If an object is at rest, can you conclude that there are no forces acting on it? Explain.

8. If a force is exerted on an object, is it possible for that object to be moving with constant velocity? Explain.

Note: The next two questions ask you to "explain" something. These are the first of many "explanations" you will be asked to make in this course, so it is worth describing what we mean in physics by an "explanation." A physics explanation should begin by stating a law or principle of physics and then proceed in a logical, step-by-step method to show how the phenomena is a consequence of the laws of physics. The explanation should be given in terms of the concepts of physics, such as force, acceleration, velocity, etc., and their properties. When you are done, a reader of your explanation should be able to understand how and why the phenomena you are explaining is a logical outcome of the known laws of physics. Descriptions that do not refer to the laws and principles of physics and do not utilize the concepts of physics will not be accepted as "explanations." Writing a good explanation is not easy, but it is an important skill to acquire and is a task you will be asked to do on many occasions.

9. Explain, from a *physics perspective*, why seatbelts are necessary in cars.

Workbook Chapter 5

10. A well-known magician's "trick" is to jerk a tablecloth out from beneath a setting of dishes and glasses without upsetting them. Is this really a "trick?" Give a physics explanation.

5.6     No Exercises

## 5.7 IDENTIFYING FORCES

For Questions 11 - 16, follow the six-step procedure of this section to identify and name all forces acting on the object.

11. An elevator suspended by a cable is descending at constant velocity.

12. A car on a *very* slippery icy road is sliding head-first into a snow bank, where it gently comes to rest with no one injured.

13. A compressed spring is pushing a block across a rough horizontal table.

14. A brick is falling from the roof of a three-story building.

Workbook Chapter 5

15. Block A and B are connected by a string passing over a pulley. Block B is falling and dragging block A across a frictionless table. Let block A be "the object" for analysis.

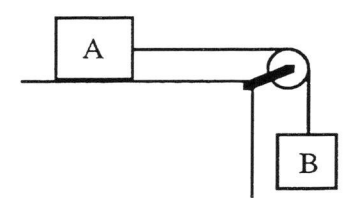

16. A jet plane is climbing at a 30° angle.

5.8     No Exercises

5.9     FREE BODY DIAGRAMS

For Questions 17 - 23:
   a) Draw a picture and identify the forces, as in Section 7, then
   b) Draw a complete free body diagram for the object, following each of the steps given in this section. Be sure to think carefully about the direction of $\vec{F}_{net}$.

**NOTE:** We will color code free body diagrams to match the colors of motion diagrams. In these and all future exercises and problems, draw *individual* force vectors with a *black* pencil and the *net* force vector $\vec{F}_{net}$ with a *red* pencil.

17. A heavy crate, to which a steel cable is attached, is being lifted straight up at constant speed by a crane.

18. A locomotive is *pushing* a boxcar along the rails at a steadily increasing speed. The friction of the rails is negligible. Consider the boxcar to be "the system."

19. A boy is pushing a box across the floor at a steadily increasing speed. Friction is *not* negligible. Consider the box to be "the system."

20. a) Your car has run out of gas while driving up a hill and is now coasting to a halt.

    b) The brakes failed, and now your car is rolling backwards down the hill.

Workbook Chapter 5

21. Betty is skiing down a hill. Despite going into a *very* strong head wind, she is still accelerating. Air resistance is *not* negligible.

22. Block B has just been released and is beginning to fall. Consider block A to be "the system."

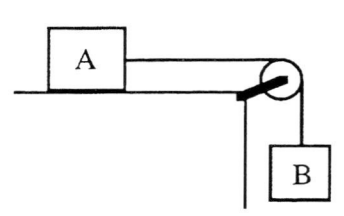

23. You flipped a coin to determine whether you or your brother has to wash the dishes. Draw separate diagrams for the coin: a) On its way up, and b) On its way back down. Air resistance is negligible.

# CHAPTER 6

# DYNAMICS I: NEWTON'S SECOND LAW

6.1  PUTTING IT ALL TOGETHER: USING NEWTON'S SECOND LAW

1. If you know all of the forces acting on an object, can you tell the direction it is moving? Explain why or why not. [Recall the comment on Workbook page 5-3 about "explanations."]

2. Write several sentences explaining why you agree or disagree with the statement: "Forces cause an object to move."

3. a) As an elevator travels *upward* at constant speed, is the tension in the cable greater than, less than, or equal to the weight of the elevator? Explain, using words, a free-body diagram, and reference to appropriate physical principles.

Workbook Chapter 6

b) As an elevator *accelerates upward*, is the tension in the cable greater than, less than, or equal to the weight of the elevator? Explain.

c) As an elevator *accelerates downward*, is the tension in the cable greater than, less than, or equal to the weight of the elevator? Explain.

6.2      No Exercises

6.3      USING THE SECOND LAW

Shown below are free body diagrams for an object with three forces being exerted on it. For each diagram, write out an algebraic statement of the $x$- and $y$-components of Newton's Second Law, Eq. 6-2. The force-side of your equations should be written in terms of the *magnitudes* of the forces $|\vec{F}_1|$, $|\vec{F}_2|$, $|\vec{F}_3|$ and any *angles* defined in the diagram.

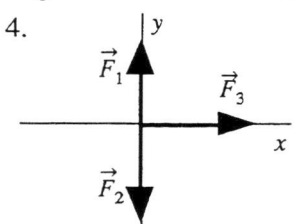

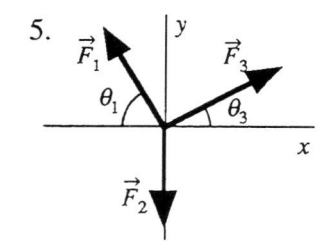

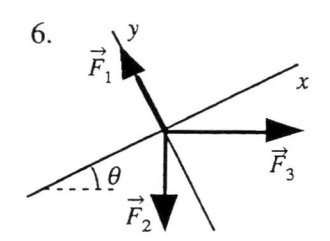

4.

5.

6.

Workbook Chapter 6

Two or more forces are exerted on a 2 kg object. They are shown below on a free body diagram that includes a grid for accurate measurement of the forces. The units of the grid are Newtons. Determine the acceleration $\vec{a}$ of the object for each case. To give your answer:
  a) Draw a vector arrow *on the grid*, starting at the origin, to show the direction of $\vec{a}$.
  [The numerical scale of the grid refers to the forces and is not relevant to $\vec{a}$.]
  b) Write the numerical value of the magnitude $|\vec{a}|$ beside the vector arrow.

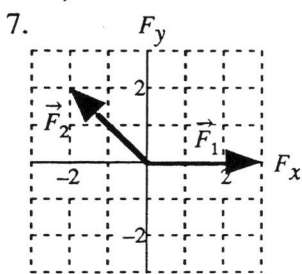

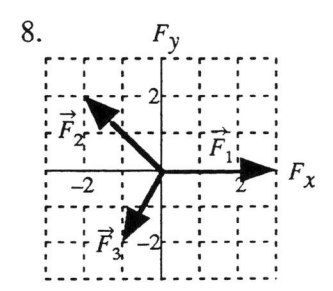

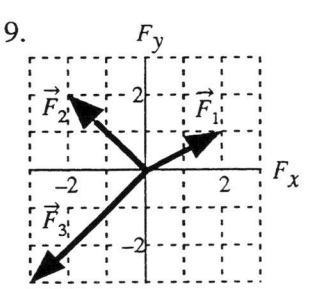

7.

8.

9.

Three forces $\vec{F}_1$, $\vec{F}_2$, and $\vec{F}_3$ are exerted, in the *xy*-plane, on a 1 kg object. Two of the forces are shown on the free body diagrams below, but the third is missing. The acceleration of the object is specified in each case. For each:
  a) Use a *black* pencil to draw (and label) *on the grid* the missing third force vector.
  b) Use a *red* pencil to draw (and label) *on the grid* the net force vector $\vec{F}_{net}$.

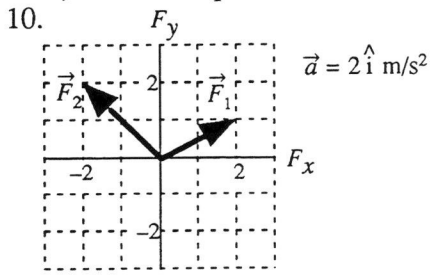

10.  $\vec{a} = 2\hat{i}$ m/s$^2$

6 - 3

Workbook Chapter 6

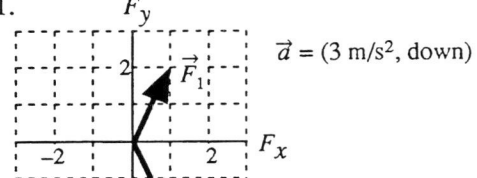

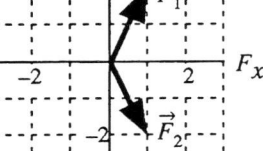

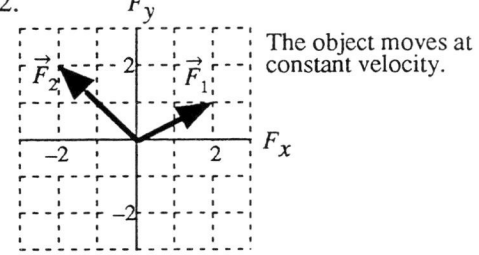

11. [Graph with $F_y$ vertical axis, $F_x$ horizontal axis, showing $\vec{F}_1$ pointing up-left and $\vec{F}_2$ pointing down-right]   $\vec{a} = (3 \text{ m/s}^2, \text{down})$

12. [Graph with $F_y$ vertical axis, $F_x$ horizontal axis, showing $\vec{F}_2$ pointing up-left and $\vec{F}_1$ pointing up-right]   The object moves at constant velocity.

## 6.4   MASS AND WEIGHT

13. Decide whether each of the following are True or False. Give a reason!
a) The mass of an object depends on its location.




b) The weight of an object depends on its location.




c) Mass and weight describe the same thing in different units.




14. a) An astronaut takes his bathroom scales to the moon and then stands on them. Is the reading of the scales his correct weight? Explain.

6 - 4

Workbook Chapter 6

b) On earth, the "conversion factor" from kilograms to pounds is 1 kg = 2.2 lb. What is the "conversion factor" from kilograms to pounds on the moon, where the acceleration due to gravity on the moon is $g = 1.6$ m/s$^2$.

15. A wood ball and a lead ball of identical shape are dropped simultaneously from a tower.
a) In the absence of air resistance, are their accelerations equal or different? If different, which has the larger acceleration? Explain.

b) Which ball hits the ground first, or are they simultaneous? Explain.

c) If, however, air resistance is present, each ball will experience the *same* air resistance force because air resistance depends only on an object's shape. In this case, are the accelerations equal or different? If not, which has the larger acceleration? Explain, using free body diagrams and Newton's Laws.

d) Which ball now hits the ground first, or are they simultaneous? Explain.

Workbook Chapter 6

16. The terms "vertical" and "horizontal" are frequently used in physics. Give *operational definitions* for these two terms. Recall, from Chapter 1, that an "operational definition" defines a term by how it is measured or determined.

## 6.5   EQUILIBRIUM

17. The vectors below show five forces that can be applied individually or in combinations to an object. Which forces or combinations of forces would cause the object to be in equilibrium?

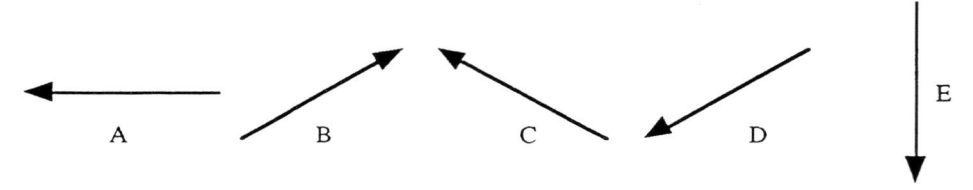

## 6.6 APPARENT WEIGHT

**18.** As an elevator *descends* it first starts from rest, then moves at steady speed, then brakes to a halt. During each of these three phases of the motion, is your apparent weight greater than, less than, or equal to your true weight? Explain, using free body diagrams.

**19.** Suppose you attempt to measure out a 100 g mass of salt, using a pan balance, while in an elevator that is accelerating upward. Will the quantity of salt be too much, too little, or the correct amount. Explain.

**20.** Suppose you have a jet-powered flying platform that can move up and down along a vertical line. For each of the following cases, is your apparent weight equal to, greater than, or less than your true weight. Explain.

a) You are ascending and speeding up.

b) You are descending and speeding up.

c) You are ascending at a constant speed.

d) You are descending at constant speed.

Workbook Chapter 6

e) You are ascending but slowing down.


f) You are descending but slowing down.


21. An box with a 75 kg passenger inside is launched straight up into the air by a giant rubber band. Once the box has left the rubber band but is still moving *upward*:
a) What is the passenger's true weight?


b) What is the passenger's apparent weight?


22. An astronaut orbiting the earth is handed two balls which have identical outward appearances. One, however, is hollow while the other is filled with lead. How might the astronaut determine which is which?

## 6.7   FRICTION

23. A block pushed along the floor with velocity $\vec{v}_0$ slides a distance $d$ after the pushing force is removed.

a) If the mass of the block is doubled but the initial velocity is not changed, what is the distance it slides before stopping?

b) If the initial velocity of the block is doubled but the mass is not changed, what is the distance is slides before stopping?

24. a) Consider a wheel *rolling* along a rough surface. Is the friction between the wheel and the surface static friction or kinetic friction? You will want to think about this one carefully.

b) To stop a car in the shortest possible distances, it is always recommended that you NOT press the brakes so hard as to lock the wheels and skid. You can stop in a shorter distance if you avoid skidding. (This is why some cars have "anti-lock brakes.") Give an explanation for this, based on your answer to part a) and what you have learned in this section about friction.

Workbook Chapter 6

25. a) Consider a crate resting on the floor of a truck. If the truck accelerates slowly, the crate has the same acceleration. What force or forces are being exerted on the crate to accelerate it? In what direction do those forces point? (You might want to draw a diagram.)

b) What happens to the crate if the truck accelerates too rapidly? Why does this happen?

# Chapter 7

# Dynamics II: Motion in a Plane

## 7.1 Trajectories

1. Complete the motion diagram for this trajectory, showing velocity and acceleration vectors.

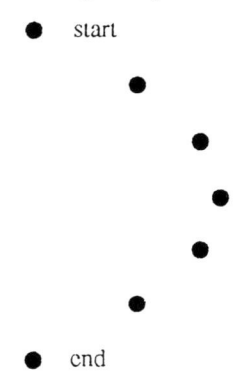

● start

●

●

●

●

●

● end

2. Shown below are the *x*-versus-*t* graph and the *y*-versus-*t* graph for a particle moving along a trajectory in the *xy*-plane.
a) Use the empty grid provided (next page) to draw a *y*-versus-*x* graph of the trajectory.

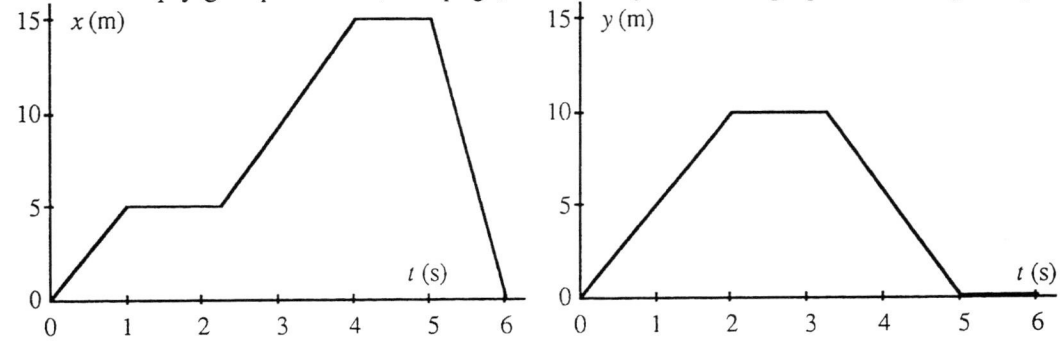

b) Draw the velocity vector at $t = 3.5$ s on your graph.
c) What is the particle's speed at $t = 3.5$ s?

Workbook Chapter 7

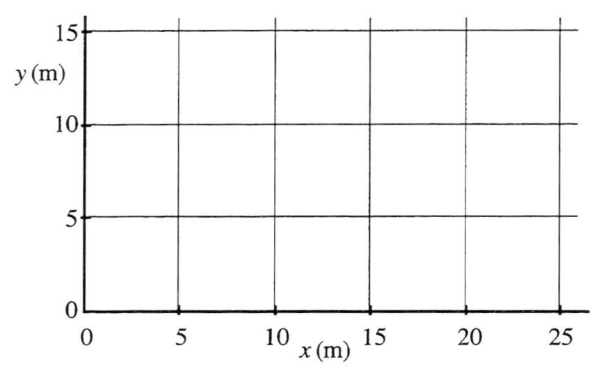

d) What is the particle's acceleration at $t = 3.5$ s?


3. Shown below is the trajectory of a particle. The particle's position at 1 second intervals is indicated with dots. The particle moves between each pair of dots at constant speed.

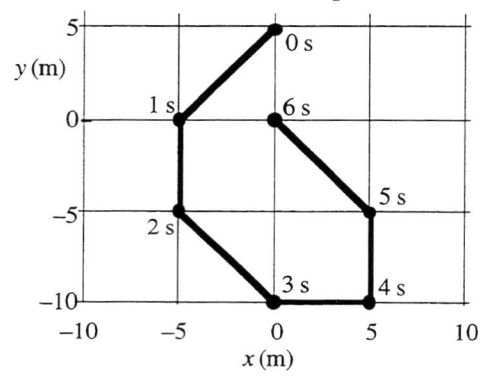

a) Use the two empty grids to draw x-versus-t and y-versus-t graphs.

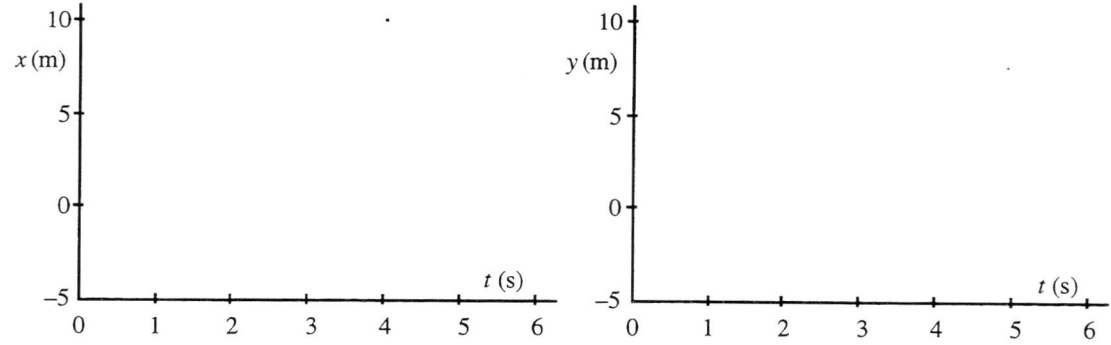

b) Is the particle's speed between $t = 5$ s and $t = 6$ s greater than, less then, or equal to its speed between $t = 1$ s and $t = 2$ s? Explain

7 - 2

Workbook Chapter 7

4. The figure shows a ramp and a ball that rolls along the ramp. Use a vector arrow to show the acceleration at each of the lettered points A to E (or write **a** = 0, if appropriate).

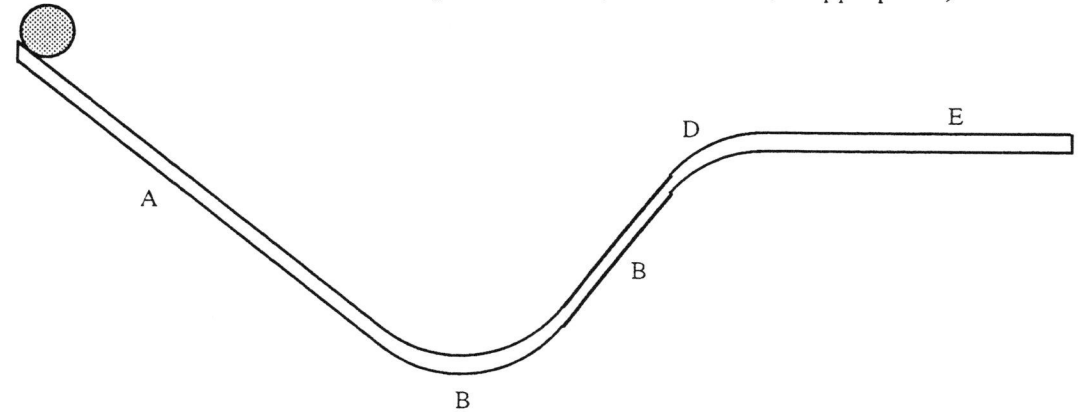

5. If you know the position vectors of a particle at two points on its trajectory and also know the time it took to get from the first point to the second, can you determine the particle's average velocity? Its instantaneous velocity? Its acceleration? Explain.

6. A rocket motor is taped to a hockey puck, oriented so that the thrust will be to the left, and the puck is given a push across frictionless ice as shown in the bird's-eye view. The rocket will be turned on by remote control as the puck crosses dotted line #2, then turned off as it crosses dotted line #3. Sketch the puck's trajectory from line #1 until it crosses line #4.

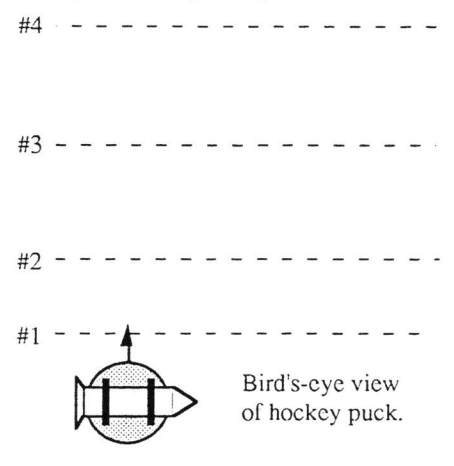

Bird's-eye view of hockey puck.

Workbook Chapter 7

7. The same puck, but now without the rocket, is again pushed across the ice in the same direction. This time it receives a sharp, very short kick to the right as it crosses line #2. It receives a second kick, of equal strength and duration but directed toward the left, as it crosses line #3. Sketch the puck's trajectory from line #1 until it crosses line #4.

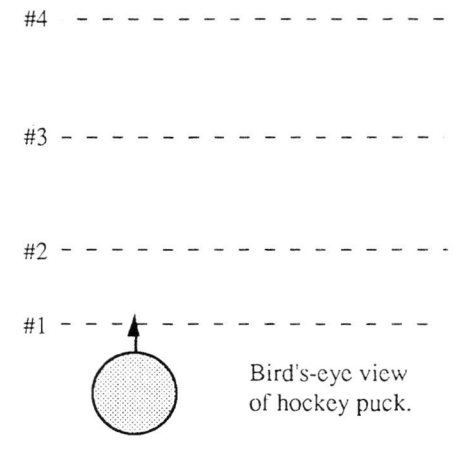

#4 – – – – – – – – – – – – – – – –

#3 – – – – – – – – – – – – – – – –

#2 – – – – – – – – – – – – – – – –

#1 – – – – – – – – – – – – – – – –

Bird's-eye view of hockey puck.

8. Tarzan swings through the jungle while hanging from a vine.

a) Use a motion diagram, as in Chapter 1, to determine the direction of Tarzan's acceleration vector $\vec{a}$
   i) Immediately after stepping off the branch, and
   ii) At the exact bottom of his swing.

b) At each of these two points, is the magnitude of the tension in the vine $|\vec{T}|$ greater than, less than, or equal to Tarzan's weight? Explain, using Newton's Laws. [Hint: Use a coordinate system at each point where $\vec{a}$ points along one axis.]

Workbook Chapter 7

## 7.2 PROJECTILE MOTION

9. As a projectile moves along a parabolic trajectory
a) Is there a point where $\vec{v}$ and $\vec{a}$ are parallel to each other? If so, where?

b) Is there a point where $\vec{v}$ and $\vec{a}$ are perpendicular to each other? If so, where?

c) Which of the following remain constant throughout the trajectory: $|\vec{r}|$, $x$, $y$, $|\vec{v}|$, $v_x$, $v_y$, $\vec{a}$?

10. The figure shows a ball that rolls down a quarter-circle ramp, then off a cliff. Sketch the ball's trajectory from when it is released until it hits the ground.

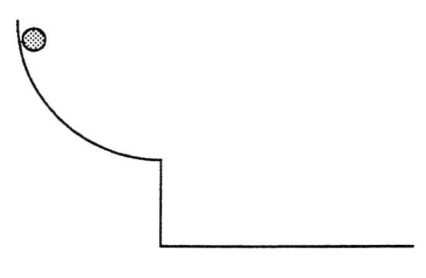

11. a) A cart that is rolling along at constant velocity fires a ball straight up. When the ball returns, will it land in front of the launching tube, behind it, or directly in it? Explain.

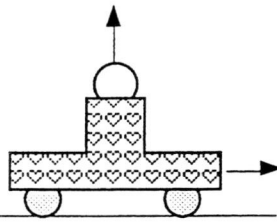

b) If the cart is accelerating in the forward direction, will your answer change? If so, how?

7 - 5

Workbook Chapter 7

12. Five balls are released simultaneously from the same height $h$ above the ground. Balls 1 - 4 all have the same initial *speed* but are launched at, respectively, angles of 45°, 0°, –45°, and –90°. Ball 5 is released from rest. What is the *order*, from first to last, with which they hit the ground? [Some may be simultaneous.]

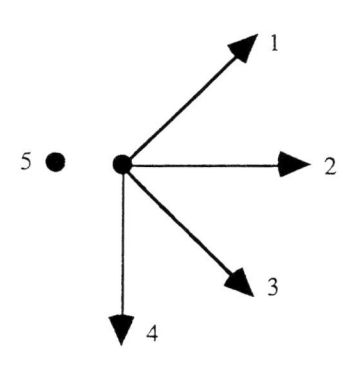

13. A rock is thrown downward, from a bridge, at 30°.

a) Sketch the trajectory on the figure.

b) Immediately after the rock is released, is the magnitude of its acceleration greater than, less than, or equal to $g$? Explain.

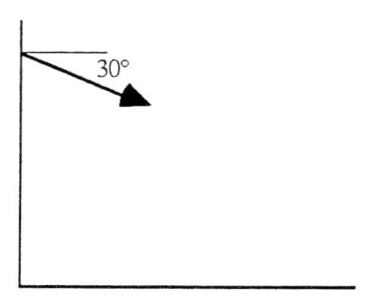

c) At the instant of impact is its speed greater than, less than, or equal to the speed with which it was released? Explain.

## 7.3 CIRCULAR MOTION AND CENTRIPETAL FORCE

14. Find the frequencies, in revolutions per second, of these circular motions:
   a) The moon orbiting the earth ($T = 27.3$ days).

   b) The earth rotating on its axis.

   c) The earth orbiting the sun.

15. The radius of the earth's orbit around the sun is 150 million kilometers. What is the speed of the earth in its orbit? Answer both in m/s and in mph.

16. a) The crankshaft in your car rotates at 3000 rpm. What is the frequency in revolutions per second?

   b) A record turntable rotates at 33.3 rpm. What is the period in seconds?

Workbook Chapter 7

17. a) The figure on the left shows a *top view* of a plastic tube that is fixed on a *horizontal* table top. A marble is shot into the tube at A with high speed. Sketch its trajectory after it passes B.

b) The figure on the right shows a ball that is swung in a *vertical* circle on a string. During one revolution, a very sharp knife is used to cut the string at the instant it is hanging vertically down. Sketch the subsequent trajectory of the ball until it hits the ground.

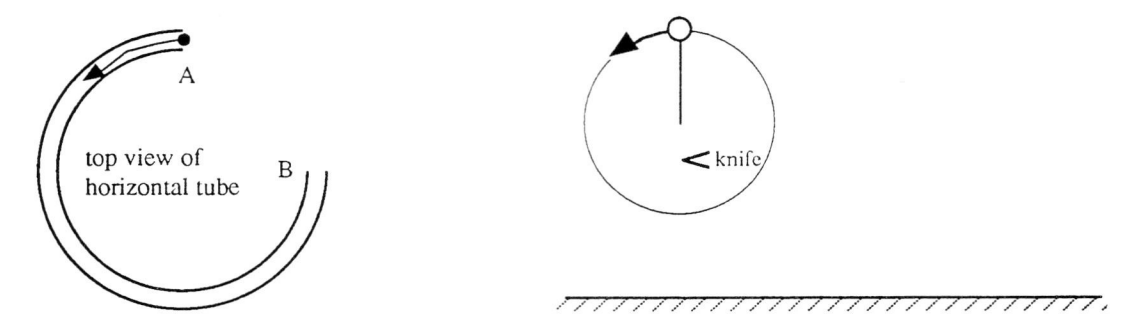

18. a) A jet airplane is flying on a level course at *constant* velocity. Draw a picture, and identify all of the forces acting on the plane and their directions. Then draw a free body diagram

b) What is the *net* force on the plane? Why? Is this consistent with your free body diagram?

c) Airplanes "bank" when they turn. Why do they do this? As always, give your explanation in terms of forces and physical laws. [Hint: What would a free body diagram look like to an observer *behind* the plane?]

## 7.4 EFFECTIVE WEIGHT AGAIN

19. A stunt plane does a vertical loop-the-loop. At which point in the circle does the pilot feel the heaviest? Explain.

20. a) You can whirl a bucket of water in a *vertical* circle, passing over your head, without spilling any as long as you whirl it fast enough. What keeps the water in the bucket? Give as complete an explanation as you can, making reference to forces and physical laws.

b) Describe in words the "threshold condition" where the water just begins to spill out of the bucket at the top of the circle?

Workbook Chapter 7

## 7.5 A NONEXISTENT FORCE

21. Why does mud fly off of a rapidly spinning car tire?

## 7.6 ORBITS AND ATOMS

22. A small projectile is launched parallel to the ground at height $h = 1$ m with sufficient speed to orbit a completely smooth, airless planet. A "bugonaut" rides in a small hole inside. Is the bug "weightless?"

# Chapter 8

# Dynamics III: Newton's Third Law

## 8.1 Interacting Systems

For Questions 1 - 5:
    a) Draw a picture showing each relevant object <u>separate</u> from all other objects, but in the correct spatial orientation. Include the earth and, if appropriate, any surfaces.
    b) Identify *all* forces and show them as <u>black</u> or <u>blue</u> vectors on the objects. Label each vector as the appropriate $\vec{F}_{\text{A on B}}$.
    c) Connect all action/reaction pairs with <u>red</u> dotted lines.

Your pictures should look like Fig. 8-2 and 8-3.

1. a) A bat is hitting a ball. (Draw you picture from the perspective of someone seeing the *end* of the bat at the moment it strikes the ball.)

   b) The ball then sails through the air.

2. a) A ball is held in your hand,

Workbook Chapter 8

    b) then is released to fall,

    c) then bounces as it hits the ground. (Consider the instant of contact with the ground.)

3. A boy pulls a wagon by a rope attached to the front of the wagon. The rope is parallel to the ground.

4. A bicycle accelerates forward from rest. (Treat the bicycle and its rider as a single object.)

5. a) A crate is in the back of a truck as the truck accelerates forward. (Treat the crate and the truck as separate systems.)




b) What specific force is responsible for accelerating the crate in the forward direction?



6. You are in the middle of a frozen lake with a surface so slippery ($\mu_s = \mu_k = 0$) that you cannot walk. It is nearly dark. You happen to have several rocks in your pocket. The ice is extremely hard, so it cannot be chipped, and the rocks slip on it just as much as your feet do. Can you think of a way to get to shore? Use pictures, forces, and Newton's Laws to explain your reasoning.




7. How do you paddle a canoe in the forward direction? As part of your explanation, in addition to words, draw pictures showing forces on the water and forces on the paddle.

Workbook Chapter 8

8. When you blow up a balloon and release it, it shoots forward. Why? Give an explanation, using both words and pictures showing forces on the balloon and on the parcel of air that was just released from the balloon.

9. How does a rocket take off? What is the upward force on it? Give an explanation, using both words and pictures showing forces on the rocket and on the parcel of hot gas that was just released from the rocket.

10. How do basketball players jump straight up in the air? As in the last few questions, give an explanation in terms of forces.

## 8.2 NEWTON'S THIRD LAW

11. Block A is pushed across a horizontal surface by someone's hand, which exerts force $\vec{F}_{\text{H on A}}$. The surface does have friction. The block is moving at constant speed.

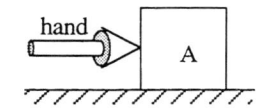

a) Draw two free body diagrams, one of the hand and the other of the block. Show only the horizontal forces, such as was done in Fig. 11 of the text. Connect action/reaction pairs. Label force vectors using the form $\vec{F}_{\text{C on D}}$. Make sure vector lengths correctly describe the relative magnitudes of the forces.

b) Rank order the horizontal forces shown in a) on the basis of their magnitudes, from the largest to the smallest. For example, if $|\vec{F}_{\text{C on D}}|$ is the largest of three forces while $|\vec{F}_{\text{D on C}}|$ and $|\vec{F}_{\text{D on E}}|$ are smaller but equal, you can record this as $F_{\text{C on D}} > F_{\text{D on C}} = F_{\text{D on E}}$.

c) Repeat both part a) and part b) for the case that the block is speeding up.

12. A second block B is added to the one of question 11, with $m_B > m_A$.

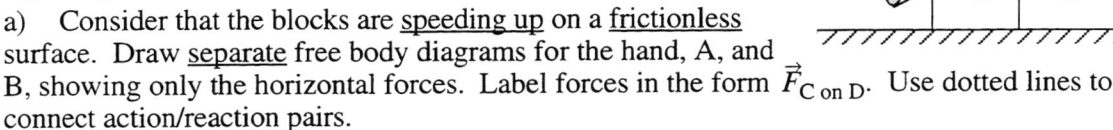

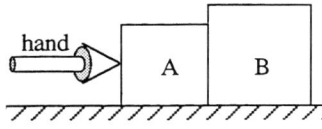

a) Consider that the blocks are speeding up on a frictionless surface. Draw separate free body diagrams for the hand, A, and B, showing only the horizontal forces. Label forces in the form $\vec{F}_{\text{C on D}}$. Use dotted lines to connect action/reaction pairs.

Workbook Chapter 8

b) By applying the Second Law to each block and the Third Law to each action/reaction pair, rank order all of the horizontal forces, from largest to smallest. Show your reasoning. Don't forget that A and B have different masses but equal accelerations.

c) Repeat a) and c) for the case that the surface has friction.

13. Blocks A and B are held on the palm of your outstretched hand and your hand is lifting them vertically at constant speed. Assume $m_B > m_A$.
a) Draw separate free body diagrams for your hand, A, and B. Show all vertical forces. Make sure vector lengths are appropriate. Connect action/reaction pairs with dotted lines. Label forces in the form $\vec{F}_{\text{C on D}}$.

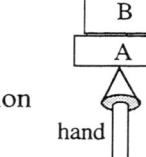

b) Rank order all of the vertical forces, from the largest to the smallest. Explain your reasoning.

c)  Suppose A and B are welded together into a single block of mass $(m_A + m_B)$. How does the magnitude of force $\vec{F}_{\text{H on A}}$ compare to the magnitude of the total weight $\vec{W}_{A+B}$? Explain.




14. A red car and a blue car, traveling with equal speeds, collide head-on and come to rest. How does the force the red car exerts on the blue car compare to the force that the blue car exerts on the red car? Are they equal, or is one larger than the other? Explain your reasoning.




15. A mosquito collides head-on with a car traveling 60 mph.

a)  How do you think the force that the car exerts on the mosquito compares to the force that the mosquito exerts on the car? Why?




b)  Draw <u>separate</u> free body diagrams of the car and the mosquito at the moment of collision, showing only horizontal forces. Label forces in the form $\vec{F}_{\text{C on D}}$. Identify action/reaction pairs.




c)  Does your answer to b) confirm your answer to a), or do you want to reconsider your answer to a)? If it confirms your answer, explain how.

# Workbook Chapter 8

## 8.3 ACCELERATION CONSTRAINTS

Write the acceleration constraint, in terms of *components*, for each of the situations shown. That is, write $(a_1)_x = (a_2)_x$, if that is the appropriate answer, rather than $\vec{a}_1 = \vec{a}_2$.

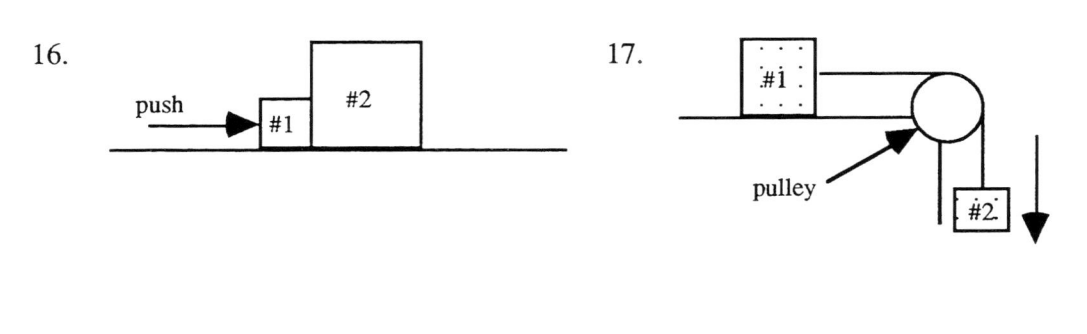

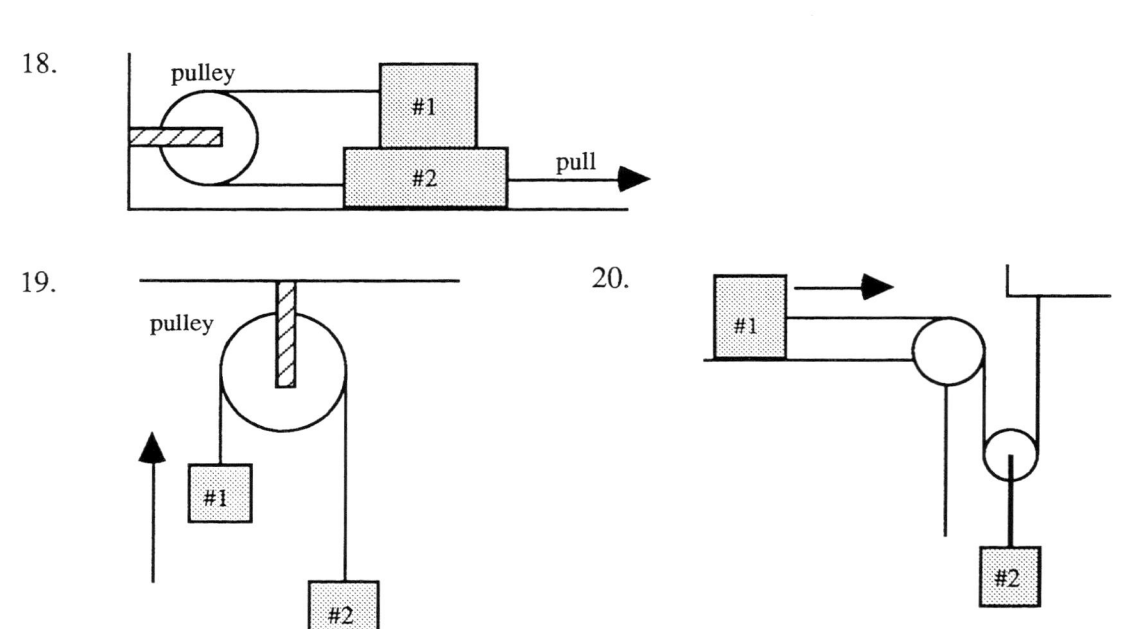

16.

17.

18.

19.

20.

## 8.4  STRINGS AND PULLEYS

For questions 21 - 26, all the masses are at rest, the strings and pulleys are massless, and the pulleys are frictionless. Determine the reading of the spring scale in each question.

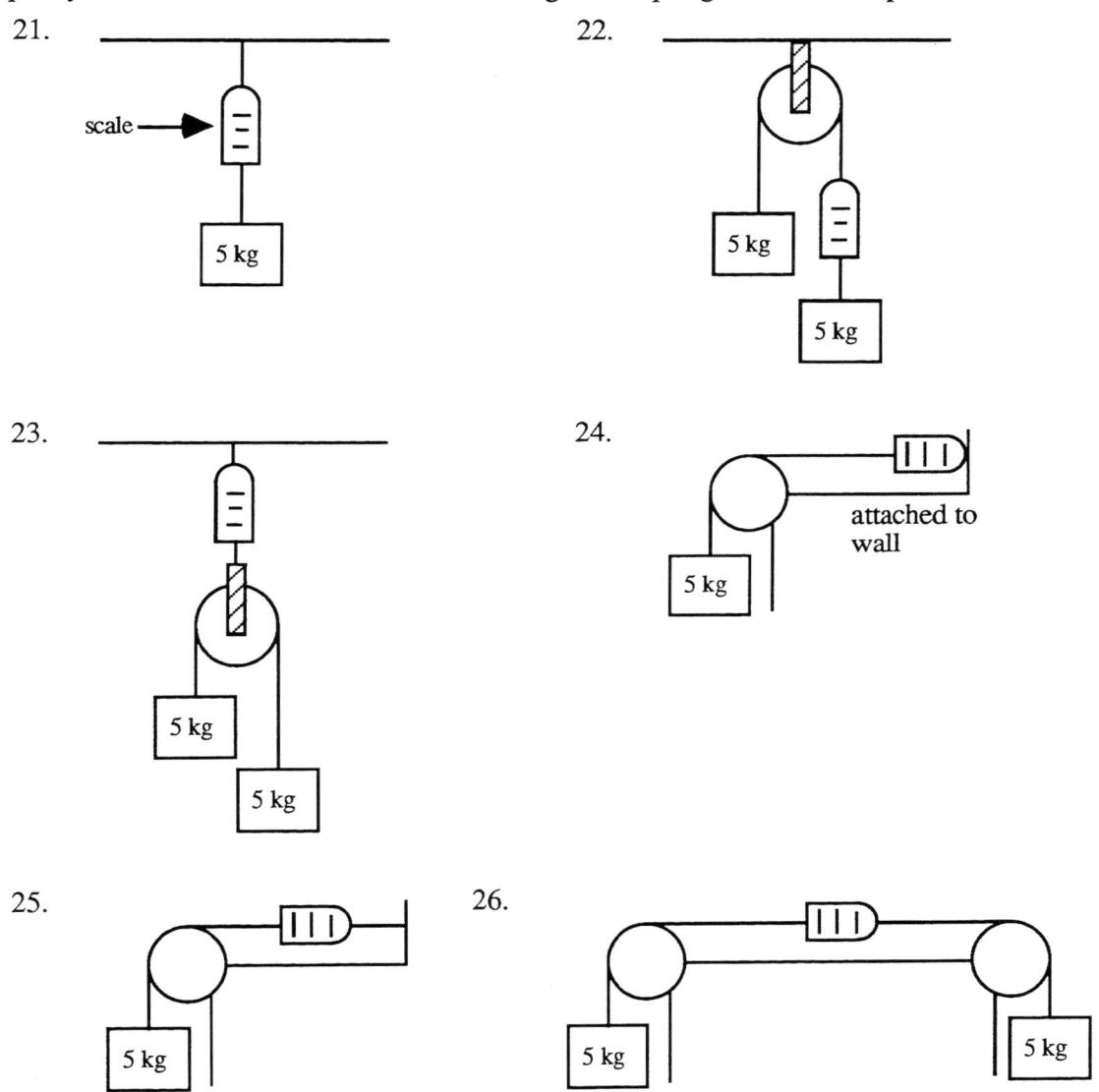

21.

22.

23.

24.

25.

26.

27. Define, in your own words, what we mean by the "tension" in a string. You should phrase your definition in terms of forces, and you likely will want to include a diagram.

Workbook Chapter 8

28. a) A tight-rope walker at the circus steps onto the high wire, causing it to sag slightly. Is the tension in the wire less than, greater than, or equal to the performer's weight? Explain, making use of a free body diagram.

b) The leading circus magazine advertises a new wire made of a material called DreamRope. The ad says that a DreamRope wire will remain perfectly straight and horizontal as the performer walks across. Should you order some? Explain.

8.5    No Exercises

8.6    EXAMPLES

29. Block's A and B, where $m_B > m_A$, are connected by a string. A hand pushing on the back of A accelerates them along a frictionless surface. The string (S) is *not* massless.

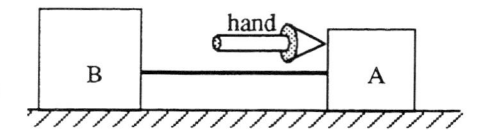

a)  Draw separate free body diagrams for A, S, and B, showing only horizontal forces. Be sure vector lengths are appropriate. Connect any action/reaction pairs with dotted lines.

b)  Using Newton's Second and Third Laws, rank order all of the horizontal forces from the largest to the smallest. Explain your reasoning.

c)  Repeat a) and b) if the string is now massless.




d)  Since $m_B > m_A$, we might expect to find $F_{S \text{ on } B} > F_{H \text{ on } A}$. Did you find this in b) and c)? Explain why this is or is not a correct statement.




30. Blocks A and B are connected by a massless string over a massless, frictionless pulley. They had been held in position but have just this instant been released. 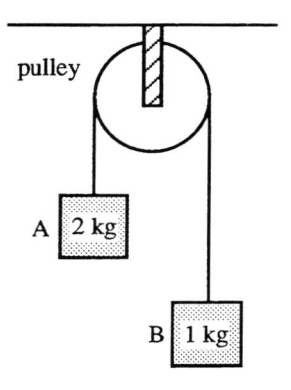
a)  Will the blocks accelerate? If so, in which directions?



b)  Draw a separate free body diagram for each block. Make vectors the appropriate lengths and connect any action/reaction pairs or "as if" action/reaction pairs with dotted lines.




c)  Rank order all of the vertical forces. Explain your reasoning.




d)  Compare the magnitude of the net force on A with the net force on B. Are they equal, or is one larger than the other? Explain.

8 - 11

Workbook Chapter 8

e) Are your answers to c) and d) consistent with your vector lengths in b)? _____
f) Is the acceleration of the block that falls less than, greater than, or equal to $g$? Explain.

31. A man pulls a crate, having $m_C < m_M$, across the floor with a *massless* rope. The floor surface does have friction -- since otherwise the man could not walk! Assume the rope is parallel to the ground.
a) Draw separate free body diagrams of the man and the crate, showing only horizontal forces. Connect any action/reaction pairs or "as if" action/reaction pairs with dotted lines.

b) How does the magnitude of $\vec{F}_{M \text{ on } C}$ compare to that of $\vec{F}_{C \text{ on } M}$? Explain.

c) Is there a net force on the crate? If so, in which direction?

d) Is there a net force on the crate? If so, in which direction?

e) If you answered Yes in c) and d), which net force is larger? Why?

8 - 12

32. A very smart three-year-old child is given a wagon for her birthday. She refuses to use it. "After all," she says, "Newton's Third Law says that no matter how hard I pull, the wagon will exert an equal but opposite force. So I will never be able to get it to move forward." What would you say to her in reply?

33. Will hanging a magnet in front of an iron cart make it go? Explain why or why not.

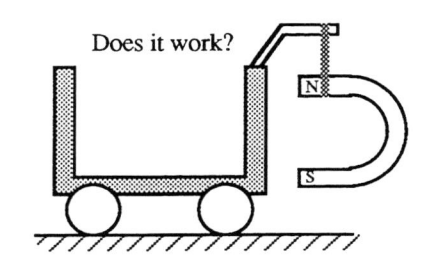

34. In Case a, Block A is accelerated across a frictionless table by a hanging 10 N weight. In Case b, the same Block A is accelerated by a steady 10 N tension in the string.

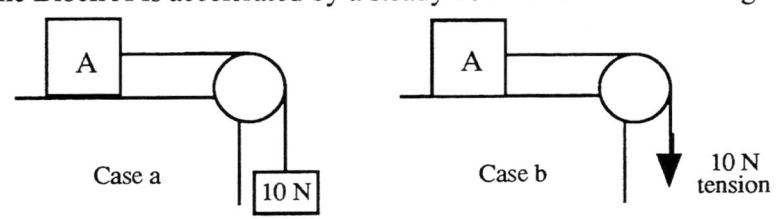

Is Block A's acceleration in Case b greater than, less than, or equal to its acceleration in Case a? Explain.

Workbook Chapter 8

For 35 - 36, draw separate free body diagrams for objects #1 and #2. Connect any action/reaction pairs (or forces that act "as if" they are action/reaction pairs) together with dotted lines.

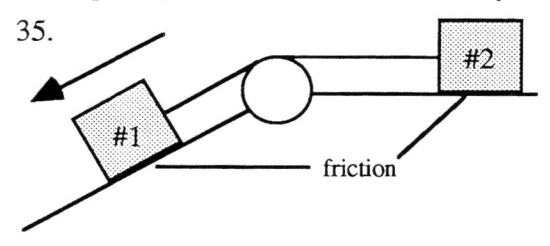

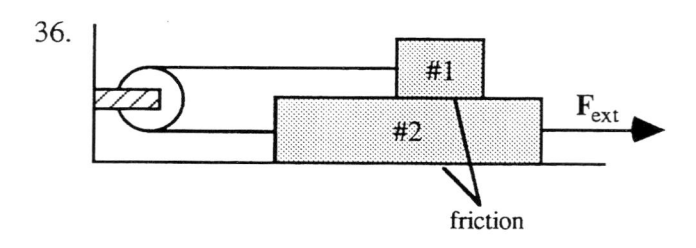

35.

36.

8 - 14

# CHAPTER 9

# MOMENTUM AND ITS CONSERVATION

9.1     No Exercises

9.2     IMPULSE AND MOMENTUM

1. Explain the concept of "impulse" in non-mathematical language. (That is, do not simply put an equation in words and say that "impulse is the time-integral of force." Explain it in terms of *physical* concepts.)

2. A 2 kg object is moving to the right with a speed of 1 m/s when it suddenly experiences an impulse, shown in a) - d). What is its speed and direction after the impulse?

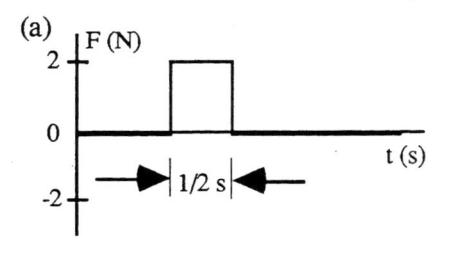

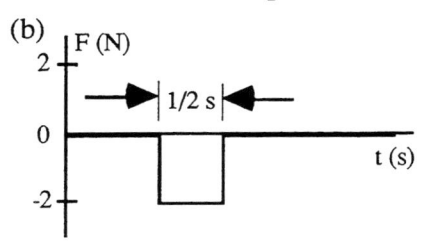

Workbook Chapter 9

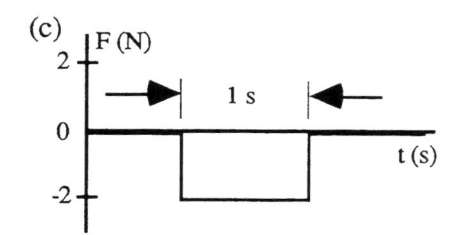

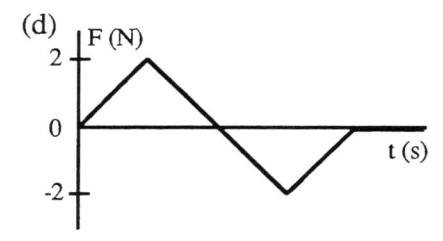

(c) Graph of F(N) vs t(s): rectangular pulse at F = -2 N with duration 1 s.

(d) Graph of F(N) vs t(s): triangular pulse rising to +2 N then falling to -2 N and back to 0.

3. For the following situations, describe *in words and pictures* what happens using 1) the language of force-acceleration-action/reaction, and 2) the language of impulse-momentum.

a) A blob of clay is thrown at a stationary bowling ball.

b) A falling rubber ball bounces off the floor.

c) Two equal masses are pushed apart by a compressed spring between them.

## 9.3  TWO PARTICLE COLLISIONS

4. A small, light ball S and a large, heavy ball L move toward each other, as shown, and undergo an elastic collision.

a) How does the force that S exerts on L compare (larger, smaller, or equal) to the force that L exerts on S? Explain.

b) How does the time interval during which S experiences a force compare to the time interval during which L experiences a force?

c) Sketch a graph showing a *plausible* $F_{L \text{ on } S}$ as a function of time and another graph showing a plausible $F_{S \text{ on } L}$ as a function of time. Be sure think about the *sign* of each force.

Force on S

Force on L

Workbook Chapter 9

d) How does the impulse delivered to S compare to the impulse delivered to L? Explain.

e) How does the momentum change of S compare to the momentum change of L? Explain.

f) How does the velocity change of S compare to the velocity change of L?

g) What is the change in the total momentum of the system -- positive, negative, or zero?

5. Two particles collide, one of which was initially at rest.
a) Is it possible for both particles to be at rest after the collision? Explain or give an example.

b) Is it possible for one particle to be at rest after the collision? Explain or give and example.

## 9.4 CONSERVATION OF MOMENTUM

6. Explain the concept of "isolated system" in non-mathematical language.

7. Explain the concept of "total external force" in non-mathematical language.

8. A golf club continues forward after hitting the ball. Is momentum be conserved? Explain, making sure you are careful to identify the "system."

9. As you release a ball, it falls -- gaining speed and momentum. Is momentum conserved?
a) Describe this situation from the perspective of choosing the ball alone as "the system."

b) Describe this situation from the perspective of choosing ball+earth as "the system."

Workbook Chapter 9

10. Just before the ball bounces its momentum is directed downward, just after the bounce it is directed upward. If the bounce is thought of as a collision, why does it appear that momentum is not being conserved?

Prepare a Pictorial Model for Questions 11 - 13, but do not solve them. Your Pictorial Models should include sketches of "before" and "after," should define symbols relevant to the problem, should list known information, and should identify the desired unknown. If this is a "two-part" problem, with a subsequent dynamics problem, your Pictorial Model should include all relevant information for both parts.

11. A 50 kg archer, standing on frictionless ice, shoots a 100 g arrow at a speed of 100 m/s. What is the "recoil velocity" of the archer?

12. The parking brake on a 2500 lb Cadillac has failed, and it is rolling slowly, at 1 mph, toward a group of small innocent children. As you see the situation, you realize there is just time for you to drive your 1000 lb Volkswagon head-on into the Cadillac and thus to save the children. With what speed should you impact the Cadillac to just bring it to a halt?

13. Fred Fingers, 60 kg, is running upfield with the football at a speed of 6 m/s when he is met head-on by Brutus, 120 kg, who is moving a speed of 4 m/s. Brutus grabs Fred in a tight grip, and they fall to the ground. Which way do they slide, and how far? The coefficient of kinetic friction between football uniforms and Astroturf is 0.3.

## 9.5 - 9.6    No Exercises

## 9.7    THE STRUCTURE OF ATOMS

11. The text claims that the nucleus takes up only about one part in $10^{11}$ of the volume of an atom. Demonstrate that this is true *without* calculating either the atom's volume or the nucleus' volume. The radius of a nucleus is $\approx 10^{-14}$ m while the radius of an atom is $5 \times 10^{-11}$ m.

12. Write a brief summary of *how we know* that atoms have a very small, heavy positive nucleus surrounded by light, negative electrons.

Workbook Chapter 9

# CHAPTER 10

# CONCEPTS OF ENERGY

## 10.1 THE LOST PENNY

1. One month Jose has income of $3000, expenses of $2500, and he sells $300 of stocks.
a) Can you determine Jose's liquid assets at the end of the month? If so, what is $L$?




b) Can you determine the amount by which Jose's liquid assets *changed* during the month? If so, what is $\Delta L$?




2. Jose begins the month with $2000 of liquid assets and $5000 of savings. His financial activity for the month looks as follows:

| Day of Month | Activity |
|---|---|
| 1 | Receives $3000 paycheck; deposits in checking |
| 3 | Spends $500 |
| 8 | Buys a $1000 savings bond |
| 10 | Pays bills totaling $1000 |
| 15 | Receives $100 birthday present from Grandma |
| 23 | Sells $1500 of stock |
| 28 | Buys a $1200 bicycle |

a) What are Jose's liquid assets and savings at the end of the month?




b) Show that Jose's Law of Conservation of Wealth is satisfied.

Workbook Chapter 10

10.2        No Exercises

10.3        THE BASIC ENERGY MODEL

3.   Upon what basic quantity does kinetic energy depend? _____

4.   Upon what basic quantity does potential energy depend? _____

5.   What are the two primary processes by which energy can be transferred from the environment to a system?

6.   What do we mean when we say that the Law of Conservation of Energy is a scientific "hypothesis?"

7.   What is meant by an "isolated system?"

8.   a) During one month, Jose transfers money out of savings and has an income that exceeds his expenditures. Can you conclude anything about the change of his liquid assets during this month? That is, can you determine if they increase, decrease, or stay the same? Explain.

b) A process occurs in which the potential energy *decreases* while work is done by the environment *on* the system. Can you conclude anything about the change of kinetic energy?

9. a) In a different month, Jose *increases* his savings while again having an income that exceeds his expenses. Can you conclude anything about the change in his liquid assets?




b) A process occurs in which the potential energy *increases* while work is done by the environment on the system. Can you conclude anything about the change of kinetic energy?




10. If the kinetic energy of a system decreases while its potential energy does not change, what is doing work on what? That is, does the environment do work on the system or does the system do work on the environment? Explain.




## 10.4    ENERGY AND MOTION

11. The figure shows the potential energy curve of a particle. Answer the following questions.

a) What position or positions are points of stable equilibrium?


b) What position or positions are points of unstable equilibrium?

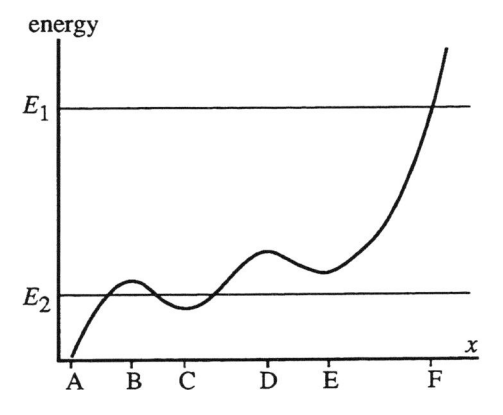

c) Suppose the particle is at position $x_A$ and moving to the right with total energy $E_1$. Describe its subsequent motion until it leaves the range of $x$ shown on the axis. You should say where the particle is speeding up, slowing down, moving at steady speed, and turning around.

Workbook Chapter 10

d) For a particle that has total energy $E_2$, what are the possible motions and where to they occur along the $x$-axis?

12. Below are a set of axes on which you are going to draw a potential energy curve. After appropriate experiments, you find the following information:
   A particle of energy $E_1$ oscillates between positions $x_D$ and $x_E$.
   A particle of energy $E_2$ oscillates between positions $x_C$ and $x_F$.
   A particle of energy $E_3$ oscillates between positions $x_B$ and $x_G$.
   A particle of energy $E_4$ enters from the right, bounces at $x_A$, then never returns.
Draw a potential energy curve consistent with this information.

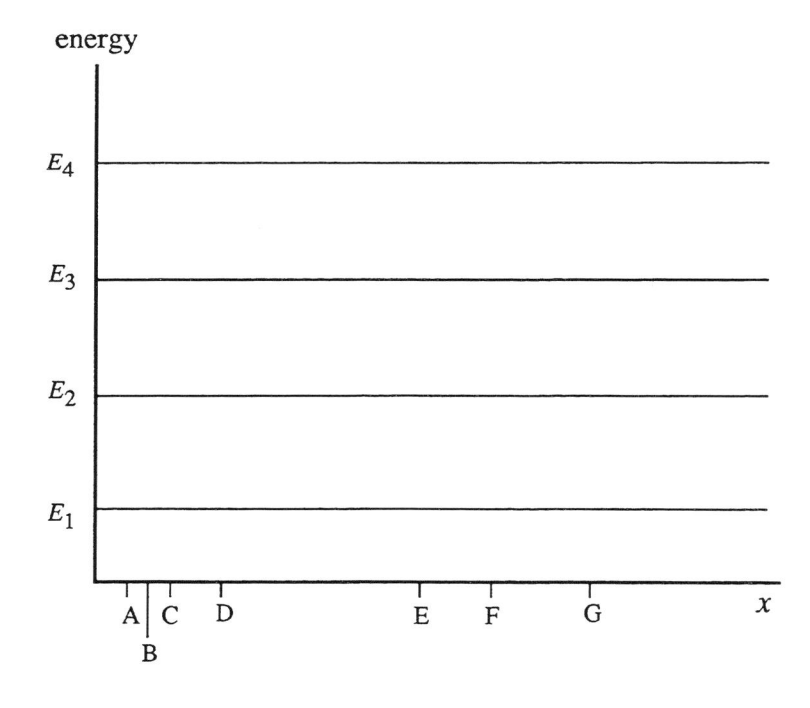

10 - 4

# CHAPTER 11

# WORK AND ENERGY

11.1    No Exercises

11.2    KINETIC ENERGY

1. Can kinetic energy ever be negative? _____
Give a plausible *reason* for your answer without making use of any formulas.




2. If a particle's velocity suddenly increases by a factor of three, by what factor does its kinetic energy change?



Particle A has half the mass and eight times the kinetic energy of particle B. How do the velocities of A and B compare?




3. On the axes below, draw graphs the kinetic energy of
a) A 1000 kg car that uniformly accelerates from 0 to 20 m/s in 20 s.
b) A 1000 kg car moving at 20 m/s that brakes to a halt with uniform deceleration in 4 s.
c) A 1000 kg car that drives once around a 40 m diameter circle at a speed of 20 m/s.
Calculate K at several times, plot the points, and draw a smooth curve between them.

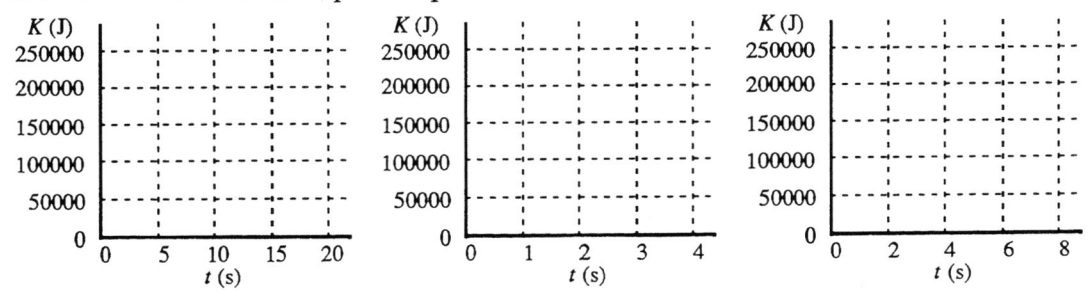

Workbook Chapter 11

## 11.3 WORK

4. For each of the situations described:
    i) Draw an appropriate diagram, similar to text Fig.2 or 5,
    ii) Identify all forces acting on the particle, and
    iii) Determine if the work done by each of these forces is positive (+), negative (−), or zero (0). Make a table beside the figure to show each force the sign of its work.

a) An elevator moves upward.

b) An elevator moves downward.

c) You push a box across a rough floor.

d) You slide down a steep hill.

e) A ball is thrown straight up. Consider it from the time it leaves your hand until the top point of its trajectory.

11 - 2

f) A ball is thrown straight up. Consider it from the top of its trajectory until you catch it on the return.

## 11.4    THE DOT PRODUCT

5. For each of the pairs of vectors shown, determine if $\vec{A} \cdot \vec{B}$ is +, –, or 0.

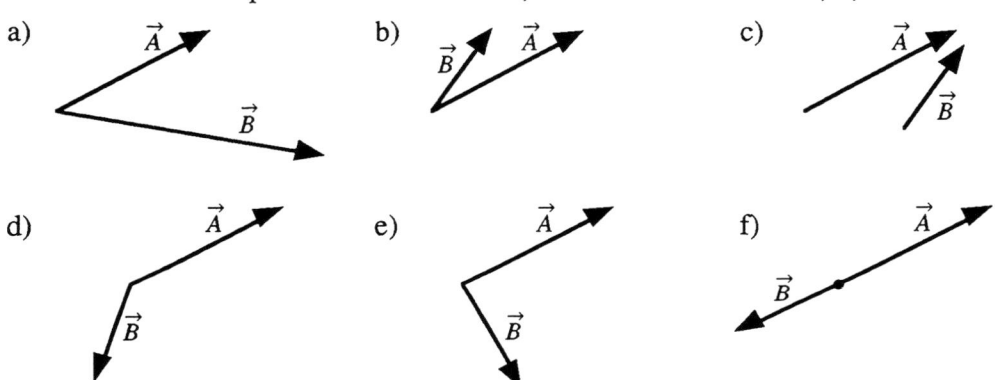

a)

b)

c)

d)

e)

f)

6. Each of the diagrams below shows a vector $\vec{A}$. Draw a vector $\vec{B}$ that will cause $\vec{A} \cdot \vec{B}$ to have the sign indicated.

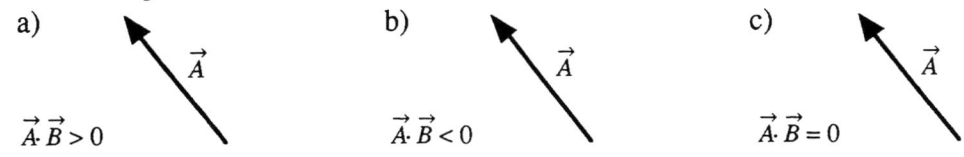

a) $\vec{A} \cdot \vec{B} > 0$

b) $\vec{A} \cdot \vec{B} < 0$

c) $\vec{A} \cdot \vec{B} = 0$

## 11.5    THE WORK-KINETIC ENERGY THEOREM

7. For each of the situations described:
    a) Draw a pictorial model, similar to text Figs. 8 -10, showing the object at the beginning (before) and end (after) of its motion. Label "before" and "after."
    b) Show and label the displacement $\Delta x$ with an arrow on the diagram.
    c) Draw a free body diagram showing all forces on the object.
    d) Make a table to show the sign (+, –, or 0) of
        i) $W$ for each force on the free body diagram,
        ii) $W_{net}$, and
        iii) $\Delta K$.

Workbook Chapter 11

a) A ball rolls to a stop along a horizontal floor with friction.

b) A ball rolls down a frictionless slope.

c) A ball rolls up a frictionless slope.

d) A ball falling after being released from rest.

e) A ball rising after being tossed straight up.

f) A descending elevator braking to a halt.

g) A rocket being launched straight up.

## 11.6   RESTORING FORCES

8. Identify three examples of elasticity, other than those mentioned in the text.




9. An elastic object is attached to the floor and pulled straight up with a string. The string's tension is measured. A graph of the tension data is shown.

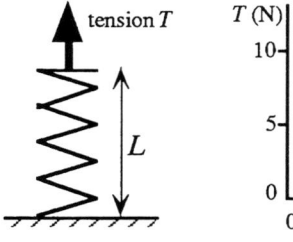

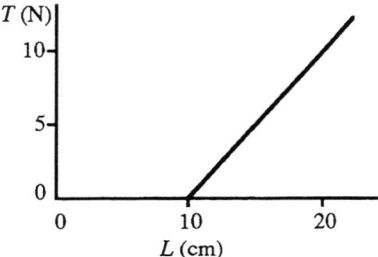

a) Does this elastic object obey Hooke's Law? Explain why or why not.



b) If it does, what is the spring constant?



c) What is $L_0$, the spring's unstretched length?



10. Draw a figure analogous to text Fig. 13 for a spring that is fixed on the right end. Use it to demonstrate that $F$ and $\Delta s$ have opposite signs for both stretching and compression.

Workbook Chapter 11

11. Bob applies a 200 N force to a spring whose left end is fixed in position, stretching it 20 cm. The same spring is then used for a tug-of-war between Bob and Bill, each pulling their end with a 200 N force.

a) How much does Bob's end of the spring move?

b) How much does Bill's end of the spring move?

c) Is your answer consistent with Hooke's Law? Explain.

12. A spring with 10 cm unstretched length exerts a restoring force $F$ when stretched to a length of 11 cm.

a) At what length is its restoring force $3F$?

b) At what compressed length is the restoring force $2F$?

## 11.7  WORK OF A VARYING FORCE

13. Show that our second definition of work, Eq. 11-10, is a *consequence* of our fourth definition of work, Eq. 11-27, if the force is constant. In other words, show that Eqs. 11-10 and 11-27 are not really separate definitions, but that Eq. 11-10 is simply a special case of the more general definition of Eq. 11-27.

14. In Example 7, the work done by the spring on the ball was found to be $W = (1/2)kx_1^2$ where $x_1$ was the initial position of the ball. In this example, $x_1 < 0$. Since, however, $x_1$ is squared, we can write $W = (1/2)kd^2$ where $d = |x_1|$ is the compression distance.

a) For $k = 10$ N/m, as in the example, calculate the work $W$ done on the ball as the spring expands for compression distances of 0, 2, 4, 6, 8, and 10 cm. Plot them on the axes below, providing an appropriate vertical scale, then draw a smooth curve through the plotted points.

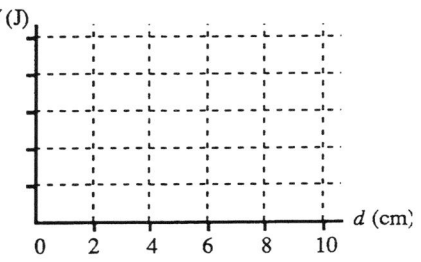

b) What geometric shape does this graph have? _____

c) Use your graph to determine how much work is done by the spring in pushing the ball through the 2 cm interval from $x = 2$ cm to $x = 0$ cm?

d) Use your graph to determine how much work is done by the spring in pushing the ball through the 2 cm interval from $x = 10$ cm to $x = 8$ cm?

e) Give an explanation as to why your answers to c) and d) are different.

15. A 0.5 kg mass on a 1 meter long string is swung in a circle on a horizontal frictionless table at a speed of 2 m/s.

a) How much work is done on the mass by the tension in the string during one revolution? Explain.

b) Is your answer to a) consistent with the work-kinetic energy theorem?

Workbook Chapter 11

## 11.8  AN ANALOGY WITH IMPULSE-MOMENTUM

16. In Chapter 9, we found a graphical interpretation of $\Delta p$ as the area under the $f$-versus-$t$ graph from an initial time $t_i$ to a final time $t_f$. Can you provide an analogous graphical interpretation of $\Delta K$, the change in kinetic energy? Explain your reasoning.

17. Return to Example 7 in the text.
a) Graph $F_{\text{spring on ball}}$ from $x_1 = -0.10$ m to $x_2 = 0$, using $k = 10$ N/m.

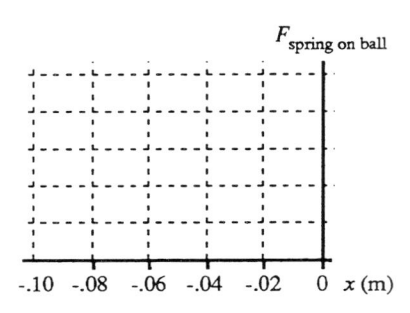

b) Find $\Delta K$ graphically, using the method you described in the previous question.

c) Use your result to b) to find $v_2$. Compare to the value found in the text.

18. A 1 kg particle moving along the $x$-axis with an initial velocity of 2 m/s at $x = 0$ is subjected to the force shown in the graph. What is the particle's velocity when it gets to $x = 5$ m?

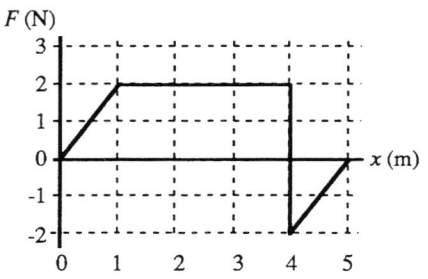

11 - 8

Workbook Chapter 11

19. Particle A, which has less mass, and particle B, which has more mass, are each pushed by equal forces through a distance of 1 m. Both start from rest.

a) Compare the amount of work done on each particle. [Here, and in subsequent questions, "compare" means to say which is larger and which smaller or, if appropriate, to say that the two values are equal.]

b) Compare the impulse delivered to each particle.

c) Compare their final velocities.

d) Compare the time it takes each particle to cover the 1 m distance.

20. Particle A, which has less mass, and particle B, which has more mass, are each pushed by equal forces for a time of 1 s. Both start from rest.

a) Compare the amount of work done on each particle.

b) Compare the impulse delivered to each particle.

c) Compare their final velocities.

d) Compare the distance traveled by each during the 1 s interval..

Workbook Chapter 11

## 11.9 POWER

21. a) If you lift a 1 kg book 1 m, how much work do you do on it?

b) How much power must you provide to lift the book in 1 s? In 10 s? In 0.1 s?

22. a) How long does it take a 60 W light bulb to use 1 J of energy?

b) With what speed would you need to lift a 2 kg mass (≈5 lb) to provide it with 60 W of lifting power?

23. In Example 13, we saw that a force responsible for constant acceleration must deliver an amount of power given by $P = ma^2t$. If a sprinter (any sprinter, not necessarily the one in the example) wants to increase her acceleration by 5% and to sustain that acceleration for the same time interval that she now does, by what percentage will she have to increase her maximum power output?

b) Does this suggest why it becomes increasingly difficult for well-trained athletes to improve their performance? Please comment.

11 - 10

# CHAPTER 12

# POTENTIAL ENERGY

12.1-12.2    No Exercises

12.3    WHAT HAPPENED TO THE WORK?

1. Describe each of the following situations from <u>both</u> a force-acceleration perspective and a work-energy perspective.

a) A falling rock.

b) Lifting a book from the floor and placing it on a shelf.

c) Pushing a plastic ball into a spring-loaded gun.

Workbook Chapter 12

2. Give a specific example of a situation in which:

a) $W_{ext} \to K$ with $\Delta U = 0$.

b) $W_{ext} \to U$ with $\Delta K = 0$.

c) $K \to U$ with $W_{ext} = 0$.

d) $U \to K$ with $W_{ext} = 0$.

e) $K \to W_{ext}$ with $\Delta U = 0$.

f) $U \to W_{ext}$ with $\Delta K = 0$.

## 12.4  GRAVITATIONAL POTENTIAL ENERGY

3. A ball rolls up an inclined plane, then back down. What is the sign of:

|  | Rolling up | Rolling Down |
|---|---|---|
| $\Delta y$ | _____ | _____ |
| $(F_{grav})_y$ | _____ | _____ |
| $W_{grav}$ | _____ | _____ |
| $\Delta U_{grav}$ | _____ | _____ |

4. A particle moves in a vertical plane along a *closed* path, starting at A and eventually returning to its starting point. How much work is done on the particle by gravity? Explain.

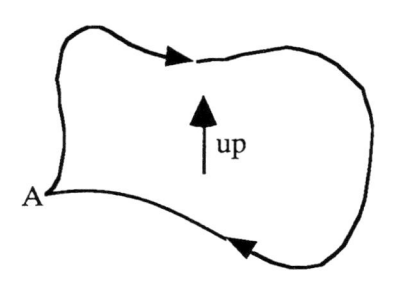

5. A roller coaster car rolls down a frictionless track, obtaining speed $v_f$ at the bottom.

a) If you want to car to go twice as fast at the bottom, by what factor must you increase the height of the hill?

b) Does your answer to a) depend on whether the track is straight or not? Explain.

## 12.5 THE ZERO OF POTENTIAL ENERGY

6. The figure below shows a 1 kg object that is initially 1 m above the ground and rises to a height of 2 m above the ground. Allan, Bill, and Charles each measure its position, but each of them uses a different coordinate system. Fill in the table to show the initial and final potential energies as well as $\Delta U$, as measured by our three aspiring scientists.

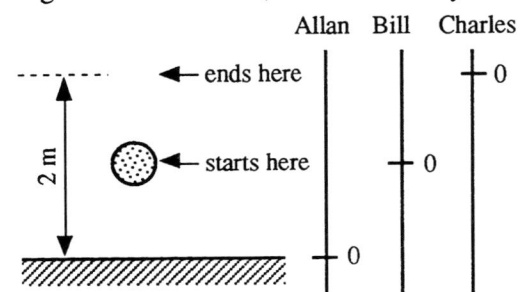

|  | $U_i$ | $U_f$ | $\Delta U$ |
|---|---|---|---|
| Allan |  |  |  |
| Bill |  |  |  |
| Charles |  |  |  |

7. Does a negative potential energy make sense? How can an object have a negative potential for converting stored energy into kinetic energy? Explain.

Workbook Chapter 12

### 12.6 CONSERVATION OF MECHANICAL ENERGY

8. In the text we have found both the work-kinetic energy theorem $W = \Delta K$ and also the definition of potential energy $\Delta U = -W$. What is the relationship between these two statements? Are they saying the same thing, or something different?

9. The text was emphatic that the conservation law is **not** $K = U$ or $\Delta K = \Delta U$. What is wrong with these two statements? Why are they not acceptable conservation laws?

10. Three balls of equal mass are fired simultaneously from the same height above the ground and with *equal* speeds. Ball 1 is fired straight up, ball 2 is fired straight down, and ball 3 is fired horizontally. Compare their speeds as they hit the ground.

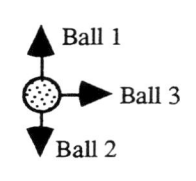

11. Below are shown three frictionless tracks. A ball is released from rest at the position shown on the left. How high does it make it on the right before reversing directions and rolling back? Point B is the same height as its starting position.

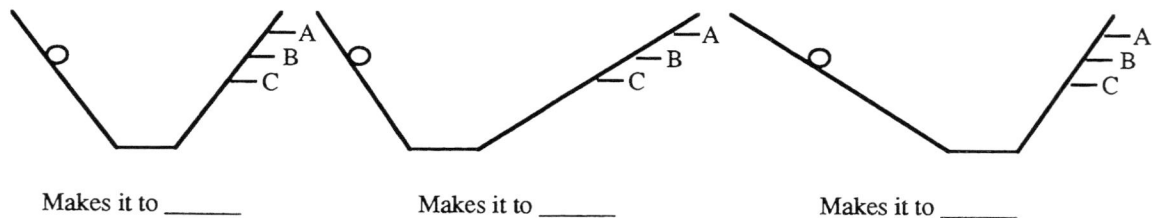

Makes it to _____      Makes it to _____      Makes it to _____

12 - 4

## 12.7  ELASTIC POTENTIAL ENERGY

12. A heavy object is released from rest at position 1 above a spring. It falls, contacting the spring at position 2 and finally coming to rest at position 3. Fill in the table below to indicate whether each of the quantities are +, –, or 0 during the intervals 1→2, 2→3, and 1→3.

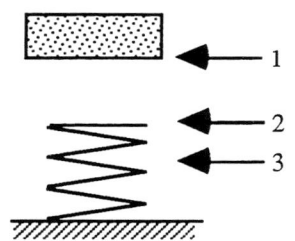

| +, –, or 0?       | 1→2 | 2→3 | 1→3 |
|-------------------|-----|-----|-----|
| $\Delta K$        |     |     |     |
| $\Delta U_{grav}$ |     |     |     |
| $\Delta U_{spring}$ |   |     |     |

13. A spring gun shoots out a plastic ball at speed $v_0$. The spring is then compressed twice the distance it was on the first shot.

a) By what factor is the spring's potential energy increased?

b) By what factor is the work needed to compress the spring increased?

c) By what factor is the ball's velocity increased?

## 12.8  No Exercises

## 12.9  POTENTIAL ENERGY DIAGRAMS

14. a) If the force on a particle is zero at some point in space, must its potential energy also be zero at that point? Explain.

Workbook Chapter 12

b) If the potential energy of a particle is zero at some point in space, must the force on it also be zero at that point? Explain.

15. When we introduced potential energy diagrams in Chapter 10, we said that the distance between the curve and the total energy line could be interpreted as the *kinetic* energy. Explain why this is so.

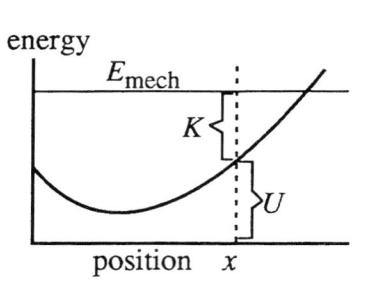

16. The graph below shows the potential energy curve of a particle moving along the *x*-axis under the influence of a conservative force.

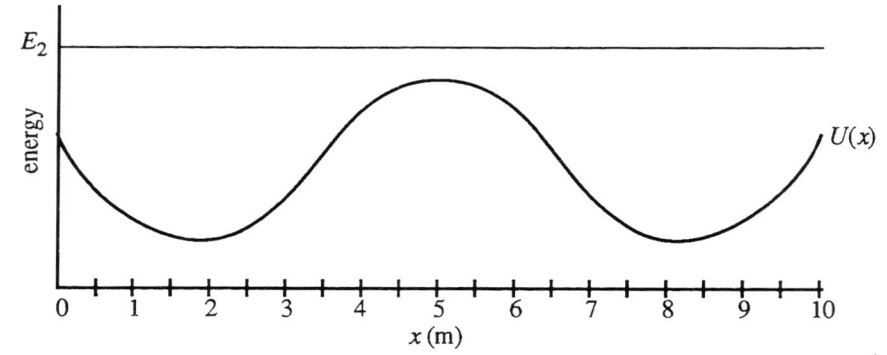

a) In which intervals of $x$ is the force on the particle to the right?

b) In which intervals of $x$ is the force on the particle to the left?

c) At what value or values of $x$ is the magnitude of the force a maximum?

Workbook Chapter 12

d) What value or values of $x$ are positions of stable equilibrium?

e) What value or values of $x$ are positions of unstable equilibrium?

f) If the particle is released from rest at $x = 0$, will it reach $x = 10$ m? Explain.

g) Draw and label on the graph a total energy line $E_1$ for a particle that undergoes oscillations about one of the stable equilibrium points you identified in d).

17. Suppose the particle in the previous question were moving to the right at $x = 0$ with total energy $E_2$. The $E_2$ total energy line is shown on the graph.

a) At what value or values of $x$ is the particle's speed a maximum?

b) At what value or values of $x$ is the particle's speed a minimum?

c) At what value or values of $x$ is the potential energy a maximum?

d) Does this particle have a turning point in the range of $x$ covered by the graph? If so, where?

e) Describe *in words* the motion of this particle. That is, describe where it is speeding up, where it is slowing down, where it stops, where it turns around, or any other important parts of its motion.

# Chapter 13

# Expanding the Concept of Energy

## 13.1　Forces That Do No Work

1. A spring of mass *m* is pushed down against the top of table, then it is released and it shoots up.

a) What force or forces are exerted on the spring?

b) Which of these forces do work and which do no work? Explain.

c) For any forces that you listed in b) as doing no work, give an example of a situation in which they *would* do actual work on the spring.

## 13.2　Microphysics and Macrophysics

2. A closed box with a few marbles inside is thrown across the room.
a) Does the box as-a-whole have kinetic energy? _____
b) Do the marbles inside have kinetic energy? _____
c) What are the similarities and the differences between the "macro" kinetic energy of the box as-a-whole and the "micro" kinetic energy of the marbles inside?

Workbook Chapter 13

13.3    No Exercises

13.4    THE CONSERVATION OF ENERGY EQUATION

3. A fire fighter slides down a fire pole at constant speed.
a) Draw a free body diagram of the fire fighter
and identify all the forces acting on him.

b) Which of these forces do work on the fire fighter? _____
c) Which of these forces can be associated with a potential energy? _____
d) What is the best choice of "the system" for using energy conservation?

e) Write down the conservation of energy equation for the system you chose in d).

f) *Interpret* your equation. What is the initial form of energy, where is it transferred to, what is the result of that energy transfer?

g) Has energy been converted into heat?

4. Consider a box full of air molecules that is accelerated to the right by an external force $\vec{F}$ across a horizontal frictionless surface.
a) Use Newton's Second Law to find an expression for $v_{cm}$ in terms of the force $F$ and the displacement $\Delta x_{cm}$.

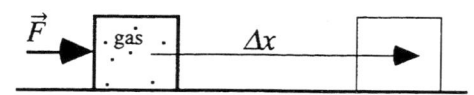

b) Using a), write an expression for $\Delta K_{cm}$.



c) Write an expression, in terms of $F$ and $\Delta x_{boundary}$, for the work done by force $\vec{F}$.



d) Using c) write the conservation of energy equation.



e) What do you conclude from b) and d) about $\Delta E_{therm}$?



f) Does this make sense? As you push the box forward, its left side is going to collide with some air molecules and knock them forward with higher speeds. They, in turn, will collide with other molecules until all of the molecules, on average, have a *higher speed* and thus a larger microscopic kinetic energy $K_{int}$. $E_{therm}$ <u>has</u> to increase. If you answered $\Delta E_{therm} = 0$ for e), then something in your analysis must be wrong. Resolve this paradox, by explaining where the error in reasoning is. [This is subtle, but important. Think carefully about the distinction between the displacement distances involved.]

Workbook Chapter 13

## 13.5  A FIRST LOOK AT THERMODYNAMICS

5. a) What is the distinction between $E_{therm}$ and $E_{int}$?




b) What is the distinction between *heat* and *work*?




c) What is the distinction between *heat* and *thermal energy*?




6. A car uses a certain amount of gasoline to accelerate from rest to 5 m/s. Compared to the gasoline used to go from 0 to 5 m/s, how much gasoline is required to go from 5 to 10 m/s? [The speeds are slow enough that neglect energy losses to air resistance.]




7. A car on level ground is pushed along by the propulsion force on its tires and is simultaneously retarded by the presence of an air resistance, or drag, force $\vec{F}_{drag}$.
a) Draw a free body diagram of the car and identify all forces acting on it.




b) Which of these forces do work on the car? _____
c) What is the best choice of "the system" for applying energy conservation?

Workbook Chapter 13

d) Write the energy equation for the system you chose in c) for an interval in which the car speeds up.

e) *Interpret* your energy equation. What are the energy transfers that occur?

For Questions 8 - 11, describe the forces and energy transfers involved in the motion. In particular: a) Identify the force or forces that accelerate or decelerate the object. b) Identify any zero-work forces. c) Identify the energy transfers involved in the motion. Your descriptions should be similar to that at the end of Example 2 (without the numbers). Pictures and diagrams will be helpful.

8. A person wearing roller blades pushing off backwards from a wall.

9. A horizontally-moving ball of putty colliding with a wall and sticking to it.

10. A woman runs up a flight of stairs.




11. A piston is pushed in to compress a cylinder of air.




13.6   No Exercises

13.7   POWER REVISITED

12. A boy on frictionless roller blades bends his arms and pushes off backwards from a wall. He puts on a lead vest that doubles his weight, then pushes off again with the same pushing force and arm bending. Compared to his first push,
a) By how much has work done on him changed?



b) By how much has his final speed (as his fingers leave the wall) changed?



c) By how much has $\Delta E_{chem}$ changed?

Workbook Chapter 13

d) By how much has the time $\Delta t$ it takes to push off changed?

e) By how much has his maximum power output changed?

13.8    No Exercises

# CHAPTER 14

# NEWTON'S THEORY OF GRAVITY

14.1 - 14.2   No Exercises

14.3   NEWTON'S LAW OF GRAVITY

1. How does the magnitude of the earth's gravitational force on the sun compare to the magnitude of the sun's gravitational force on the earth? Larger, smaller, or the same? Why?

2. In a binary star system (two stars circling each other) Star A is three times as massive as Star B.
a) Compare the magnitude of the force on Star A to the force on Star B.

b) Compare the acceleration of Star A to the acceleration of Star B.

Workbook Chapter 14

3. Comets orbit the sun in highly elliptical orbits. A new comet is sighted on Day 1.
a) On Day 30, the comet's acceleration $a_{30}$ is observed to be twice as large as its acceleration $a_1$ was on Day 1. How does the comet's distance from the sun $r_{30}$ on Day 30 compare to its distance $r_1$ on Day 1?

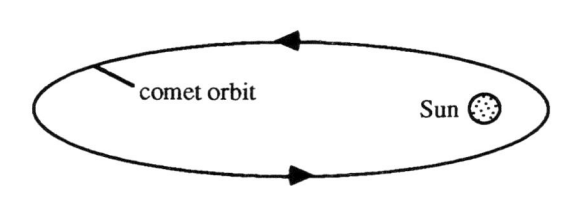

b) On Day 60 the comet has rounded the sun and is headed back out to the farthest reaches of the solar system. The force $F_{60}$ on the comet is the same on this day as the force $F_{30}$ was on Day 30, but the comet's distance from the sun $r_{60}$ is only 90% of its distance on Day 30. Astronomers recognize that the comet has lost mass -- it was "boiled away" by the heat of the sun during the time of closest approach, between Days 40 and 50, and formed the comet's tail. What percentage of its initial mass did the comet loose?

## 14.4    BIG $G$ AND LITTLE $g$

4. Explain why the Space Shuttle astronauts are "weightless."

5. How far away from the earth does an orbiting spacecraft have to be in order for the astronauts inside to be weightless?

## 14.5 WEIGHING THE EARTH

6. Don't do any calculations, but <u>describe</u> how to "weigh the sun."

## 14.6 ORBITS

7. Planet X, circling the star Omega, has a "year" that is 200 earth days long. Planet Y circles Omega at twice the distance of Planet X. How long is a year on Planet Y?

## 14.7 No Exercises

Workbook Chapter 14

## 14.8 GRAVITATIONAL AND ELECTRICAL ENERGY

8. Explain <u>why</u> the potential energy of two masses is negative. [Note: Saying "because that's what the formula gives" is <u>not</u> an explanation. An *explanation* requires going back to the basic ideas and definitions of force and potential energy.]

## 14.9 ORBITAL ENERGETICS

9. a) When the Space Shuttle wants to return to earth from a circular orbit, in which direction does it fire its rocket engine?

b) Make a drawing, similar to Fig. 15, showing the earth, the Shuttle's orbit before firing its engine, and its new orbit after firing its engine.

10. Suppose the earth had no atmosphere, so the Shuttle would continue unimpeded along its new orbit until intersecting the ground (ouch!). As it descends, would its speed increase, decrease, or stay the same? Explain your answer in terms of energy transfer.

14 - 4